"Without diluting or diverting the core message of 'R2P' – designed specifically to energize the world's response to genocide and other heinous mass atrocity crimes – Parr creatively applies its principles and methodology to meeting the existential challenge of climate change. A thoughtful, stimulating, practical, and very-readable exercise in the cross-fertilization of ideas."

Gareth Evans, *Former Australian Foreign Minister, President Emeritus of the International Crisis Group and author of 'The Responsibility to Protect'*

Climate Change Action and the Responsibility to Protect

This book brings together two important fields in the study of international politics and policy: climate change adaptation and mitigation (climate action) and the Responsibility to Protect (R2P). Both have attracted strong scholarly attention in each of their respective research silos, but there is yet to be a strong research push that explores the relationship between the two.

Filling this gap, Ben L. Parr argues that the climate action and the R2P agendas share a common goal: to protect vulnerable human populations from large-scale harm. To substantiate this argument, Parr reveals where the historical, conceptual, and operational parallels exist between the two agendas, and where and when researchers and practitioners from both camps might work together in practice to achieve their common goal in the challenging years ahead. Notably, the book builds on recent efforts by Western governments in the UK, US, and EU to integrate climate action policies into conflict prevention and response policies. To achieve this, the volume situates a variety of climate action policies alongside the 46 policy options found in the R2P operational framework (commonly known as the R2P toolbox) across its prevention, reaction, and rebuilding phases.

Climate Change Action and the Responsibility to Protect will be of significant interest to policy-orientated students and scholars, those working at the academic-policy interface in the NGO community, as well as those working in government and international organisations.

Ben L. Parr has researched climate policy and politics since 2009 and is a former staffer to key R2P architect, Professor the Hon Gareth Evans AC KC.

Routledge Advances in Climate Change Research

Climate Change in the Global Workplace
Labour, Adaptation and Resistance
Edited by Nithya Natarajan and Laurie Parsons

Governing Climate Change in Southeast Asia
Critical Perspectives
Edited by Jens Marquardt, Laurence L. Delina and Mattijs Smits

Kick-Starting Government Action against Climate Change
Effective Political Strategies
Ian Budge

The Social Aspects of Environmental and Climate Change
Institutional Dynamics Beyond a Linear Model
E. C. H. Keskitalo

Climate Change and Tourism in Southern Africa
Jarkko Saarinen, Jennifer Fitchett and Gijsbert Hoogendoorn

Climate Cultures in Europe and North America
New Formations of Environmental Knowledge and Action
Edited by Thorsten Heimann, Jamie Sommer, Margarethe Kusenbach and Gabriela Christmann

Urban Planning for Climate Change
Barbara Norman

Climate Action in Southern Africa
Implications for Climate Justice and Just Transition
Edited by Philani Moyo

For more information about this series, please visit: www.routledge.com/Routledge-Advances-in-Climate-Change-Research/book-series/RACCR

Climate Change Action and the Responsibility to Protect

A Common Cause

Ben L. Parr

Routledge
Taylor & Francis Group
LONDON AND NEW YORK

earthscan
from Routledge

First published 2024
by Routledge
4 Park Square, Milton Park, Abingdon, Oxon OX14 4RN

and by Routledge
605 Third Avenue, New York, NY 10158

Routledge is an imprint of the Taylor & Francis Group, an informa business

British Library Cataloguing-in-Publication Data
A catalogue record for this book is available from the British Library

ISBN: 978-0-367-20135-7 (hbk)
ISBN: 978-1-032-59273-2 (pbk)
ISBN: 978-0-367-20136-4 (ebk)

DOI: 10.4324/9780367201364

Typeset in Times New Roman
by Newgen Publishing UK

Contents

Preface

In 2011–2012, I was employed as the Personal and Research Assistant to the former Foreign Minister of Australia and co-creator of the Responsibility to Protect (R2P) concept, the Hon Gareth Evans AC KC. This was a period of intense learning, especially in terms of helping me understand what it took to make a difference in the world. My obsession at this time however was my doctoral research on Australian climate policy and diplomacy (with a strong focus on Australian foreign policy). But I knew that something big was happening in Gareth's world, in which I was a bit player.

I now understand that that big thing was probably the high-water mark for the R2P concept, at least in terms of its operational side. In March 2011 (two months after I had started as Assistant) a multinational force led by NATO intervened militarily in Libya to stop Muammar Gaddafi's forces from carrying out a large-scale massacre of civilians fleeing Benghazi. The R2P principle was used as a justification for the intervention. When I left Gareth's office, I thought nothing more of the R2P for quite some time.

In April and July 2016, Russian and Syrian jets and helicopters began indiscriminately bombing civilians in Aleppo. This had been happening elsewhere in the country for a while, with chemical weapons used on several occasions. But the daily videos streaming into my inbox from Aleppo of lifeless, wet, and bloodied children's bodies covered in grey cement dust was really too much to bear. So, I asked myself, if I am so heartbroken by these images, what am I going to do about it? I reached for Gareth's book *The Responsibility to Protect: Ending Mass Atrocity Crimes Once and For All*. I was immediately all-in.

More than anything I was struck by the 'operationalising R2P' chapters; here it was, actual instructions detailing what decision makers (and wannabe advocates like myself) could do, and when, to stop kids and their families from being massacred. However, as I read – at this stage (2017), working on international climate policy – I was also struck by the absence of what I understood to be the most important challenge facing humanity: climate change. As I made notes in Gareth's publication about possible climate-R2P links (and being generally aware of R2P's scarcity in the climate policy research), I thought, I am going to write a book about the relationship between the R2P and climate action. This is something I can do.

And here it is.

A Common Cause argues that the climate action and the R2P agendas share a common goal: to protect vulnerable human populations from large-scale harm. The book reveals where the historical, conceptual, and operational parallels exist, and where and when researchers and practitioners from both camps might work together in practice to achieve their common goal in the challenging years ahead.

I would like to acknowledge Gareth Evans for his inspirational work on R2P (as well as Australian foreign policy) and thank him for engaging in this book writing project and for the other support he has given me over the years. I would also like to thank Thomas G. Weiss and Robyn Eckersley for their suggestions on this book. Thanks also to Nicholas Farrelly for supporting my adjunct position at the University of Tasmania so that I may complete the book. Thanks to Routledge for their patience. But as always, the biggest thanks must go to my family, Amelia and our two boys, Luka and Aurelien, for their support, understanding, and love over the many years it has taken me to complete this protect. Thanks also to my parents and siblings, especially Sophie Parr, for their feedback on first and last drafts.

Introduction

Global warming is accelerating and mass atrocities proliferating. In August 2021, the Intergovernmental Panel on Climate Change (IPCC), the chief international body tasked with synthesising the science of climate change, released its Sixth Assessment Report (AR6) on the physical science of climate change (Working Group I).[1] It found that the since 1880 (when instrumental records began) the earth's land surface temperature has risen by 1.1 degrees Celsius (°C); nine of the past ten years have been the warmest on record; the build-up of carbon dioxide in the atmosphere has grown from 280 parts per million (ppm) to 419 ppm between 1880 and 2022; and that warming has caused an increased frequency and intensity of climate extremes including storms, floods, heatwaves, and droughts – which are becoming increasingly dangerous for human populations.[2] Researchers examining atrocities and conflict have found that since the Arab Spring in 2011 (after a period of decline in the 1990s and 2000s[3]) instances of armed conflict have increased; the frequency and scale of atrocity crimes have increased dramatically; the abuse of children in the context of armed conflict (e.g. killing, maiming, and rape) appears to have increased; 'one-sided violence', which entails the targeting of unarmed civilians by states or organised non-state groups, has also increased; and the number of displaced people due to armed violence is at an all-time high since the end of the Cold War.[4]

International attempts to curb global warming through reducing carbon emissions began in 1992 with the Rio Earth Summit and the signing of the United Nations Framework Convention on Climate Change. Since then, several major international agreements have been reached (e.g. in Kyoto in 1997, Copenhagen 2009, Paris 2015, and Glasgow 2021) aimed at preventing warming of the climate system considered 'dangerous' to humans – which in 2009 was agreed as warming that exceeded 2°C above preindustrial levels. In addition to these international efforts, some regions, countries, and subnational entities (e.g. states/provinces and cities) have established a price on carbon emissions and or developed technologies to reduce emissions in key emitting sectors – namely, energy, transport, heavy industry (manufacturing and mining), and agriculture (where methane is the major problem). Key climate adaptation

DOI: 10.4324/9780367201364-1

policies aimed at human protection – which, similarly to mitigation policies, have been negotiated though multilateral, regional, and bilateral fora – have included a variety of major climate financing initiatives (although sometimes an agreement-delivery mismatch is evident), technology transfer, and Disaster Risk Reduction programmes. But despite all these efforts, emissions, warming, and impacts have continued unabated.

Major atrocity/conflict (this relationship is clarified below) prevention and response policy initiatives have their roots in the aftermath of the Second World War with the Nuremberg trials of Nazi figures who were central to the Jewish Holocaust, the establishment of a series of UN conventions on genocide, as well as the Universal Declaration of Human Rights. The next major surge of activity occurred in the aftermath of the Cold War with successive UN secretary-generals championing major UN-driven conflict prevention analyses and initiatives, the establishment of two major international tribunals to prosecute the perpetrators of mass atrocities committed in Rwanda and the former Yugoslavia, along with a variety of conceptual contributions that challenged and reformed the idea of 'humanitarian intervention'.

One such contribution commenced with the launch of the *International Commission on Intervention and State Sovereignty* (ICISS) in September 2000. Australia's former Minister for Foreign Affairs, Gareth Evans (and a key figure in this book), co-Chaired the Commission and after 12 months of consultations across the globe, in December 2001, the Commission published its final report under the title *The Responsibility to Protect* (R2P).[5] Importantly, the report recast state sovereignty as a responsibility, not a right.[6] From this foundational change, the R2P's two basic principles emerged. First, the state itself had the primary responsibility to protect its population from four atrocity crimes (genocide, war crimes, ethnic cleansing, and crimes against humanity). Second, should the state in question manifestly fail in that responsibility – or indeed is itself the perpetrator – the responsibility to protect the population falls to the international community.[7] The report identified a variety of actions that the international community could take to protect populations, which were overwhelmingly of a preventive nature encompassing diplomatic, economic, legal, and security strategies, but also, as an absolute last resort in extreme situations, military intervention. In September 2005, the R2P was adopted unanimously by the UN General Assembly at the World Summit, in which all 193 Member States of the UN are represented.[8] In late 2010 and early 2011, as the Arab Spring unfolded, the R2P principles (and response options) provided the justification for a NATO-led military intervention in Libya that successfully saved thousands of innocent civilians from slaughter.[9] However, problems with the mission firmed up opposition to the R2P concept. As mentioned, since 2011 atrocities have proliferated globally.

Given the deeply complex and interrelated relationship between climate and conflict, which is unpacked throughout this book, Western countries (which in this book refers to the US, UK, and European Union [EU]) have increasingly sought to integrate climate change issues and policies into traditional defence and foreign policies and programmes.[10] This push commenced in a serious way in 2010. The pace quickened in the aftermath of the Paris Agreement on climate change being reached in 2015. And it accelerated again with the incoming Biden Administration in 2020. Integration efforts overall have tended to focus on identifying the risks posed by climate change to national and international security. However, in 2021 this focus changed with the EU and US publishing ground-breaking solutions-orientated reports that integrated several climate action policies into conflict prevention and resolution policies – the EU even organising its integrated policy options under multiple phases: 'conflict prevention', 'crisis response', and 'conflict resolution and stabilisation'.[11] Western countries have been more cautious when it comes to integrating climate change concerns into atrocity prevention and response frameworks. For example, the most recent frameworks from the US (2022) and UK (2022) cautiously identify climate impacts as a possible risk and driver of atrocities, but do not present any climate-informed policy solutions.[12]

The book aims to add value to this recent surge of Western activity seeking to integrate climate action policies into conflict/atrocity prevention and response policies (encompassing defence, foreign, economic, and legal domains). It suggests that while many important attempts have been made, it is the existence of deeply held *Common Cause* (and mission) about protecting vulnerable human populations from large-scale harm (shared between the climate action and R2P agendas) that provides an especially strong foundation on which to build a collaborative agenda between practitioners and researchers from both camps over the coming years.[13] Developing a cooperative agenda could be mutually beneficial. For instance, on the one hand, the climate action agenda is searching for new policies and an overarching policy framework to prevent dangerous warming of the climate system and the commensurate impacts on human populations, which the R2P agenda can most certainly provide. On the other hand, the R2P agenda is searching for something to help reenergise the concept and move it beyond its past controversies, which could be helped along by including a forward-looking climate action agenda.

Little scholarly or applied research has covered this relationship. For example, none of the 53 chapters comprising *The Oxford Handbook of The Responsibility to Protect* (2016) are devoted to climate change, the environment, or anything else related; and indeed only twice does 'climate change' feature in the 1129 pages: one briefly questioning whether climate change and HIV should be an issue of UN Security Council concern, while the other, written by former Canadian Foreign Minister, Lloyd Axworthy, and the original funder of the entire R2P enterprise, who wrote that 'considering the nexus of climate change, migration, and conflict, a more broadly conceived R2P could provide a framework for compelling a "responsibility to survive"'.[14] Axworthy has been a consistent voice since the birth of the R2P in support of integrating climate-related issues into the R2P framework (explained later in detail).[15]

Despite this, three key themes in the R2P literature are somewhat discernible, which can be captured under Risk, Scope, and Cases. First, much of the

R2P literature, should it discuss climate change, tends to focus on the relationship between climate impacts and atrocities. Overwhelmingly, this relationship is viewed through the 'threat-multiplier' lens[16] – which posits that climate impacts interact with and potentially exacerbate pre-existing security threats present in political, economic, and social domains.[17] Second, another major theme is the debate over whether the existing R2P scope – encompassing genocide, war crimes, ethnic cleansing, and crimes against humanity – should be extended to include climate-related human protection catastrophes such as major cyclones and foods.[18] Overwhelmingly, R2P authors oppose a revised scope seeing any dilution of R2P's focus on atrocities as putting at risk the hard-won consensus of 2005.[19] Third, considerable scholarly analysis and debate has been devoted to unpacking the international response to Cyclone Nargis in 2008 in which 130,000 people were killed and 2.4 million displaced in Myanmar (see below);[20] fewer still cover the environmental dimension to the Darfur atrocity situation in 2003;[21] and fewer again, cover the role that drought played in sparking Syria's civil war in 2010.[22] As we shall see, these three themes on many occasions are interlinked.[23]

In practice, the most prominent and controversial climate-R2P relationship – which has driven much of the above scholarly debate – involves the ICISS report of 2001 identifying environmental emergencies as grounds for an R2P response, including the use of military force, which was subsequently narrowed to the four-fold scope above by the time of the World Summit of 2005. The other major 'real-world' climate-related R2P situation involved the French Foreign Minister, Bernard Kouchner, who in the aftermath of Cyclone Nargis explicitly called for an R2P response involving the international community delivering aid to those suffering inside Myanmar without the government's consent.[24]

The climate policy scholarly literature tends to convey a similar general sense of apprehensiveness about drawing links between R2P and climate.[25] On the rare occasion that R2P is discussed, there is a tendency to focus on the possibility of military intervention in response to environmental emergencies[26] or other instances of major environmental damage such as an ecocide situation.[27] This focus has tended to direct criticism towards the Permanent Five members of the Security Council (the UK, the US, France, Russia, and China), which are required to authorise the use of force.[28] The criticism thus follows that the R2P and climate 'mesh poorly' because the P5 would not support a climate-related intervention because they are also the world's major carbon emitters.[29] Another key criticism is that there is no room for a climate-informed R2P, and that the dominant concept, human security, 'gets the job done'.[30] It remains unclear what that 'job' is, and as we shall see, the human security concept itself has major flaws that limit its utility – one of which is the absence of any operational content.

The Environmental Peacebuilding (EP) concept however – which has equally deep and sophisticated conceptual underpinnings as the human security concept[31] – has developed an operational component.[32] But when compared to the R2P framework, EP seems to suffer from the following.[33] First, while some contemporary definitions of the concept span three operational phases (prevention, reaction, and rebuilding),[34] EP has generally been geared towards the rebuilding phase.[35] Second, while its policy reach is considerable – although not as extensive as the R2P operational framework – options are not organised under a standardised framework or with policies that graduate in severity from less to more coercive. Third, EP tends to focus on 'cooperative solutions',[36] whereas the R2P framework encompasses not only cooperative solutions (e.g. diplomatic peacemaking and positive inducements), but also a variety of coercive options (e.g. 'gun boat diplomacy' and military intervention). Fourth, EP is generally concerned with interstate conflict and cooperation;[37] by contrast, R2P is focused on intrastate conflicts and atrocities, which is where most of the modern-day civilian casualties are found. Having said this, the EP scholarship and applied work presents the closest resemblance, and suitable conceptual springboard to develop a climate-informed R2P framework, and is drawn upon throughout the present book. Indeed, the subterranean unifying concept in this book is cooperation – a core EP tenant that refers to the need for actors and institutions to find a common ground in the interests of peace.

A positive use of the R2P framework in practice has been developed by the Washington, DC–based Centre for Climate and Security, with various reports published under the title, *The Responsibility to Prepare (and Prevent)*.[38] These collective reports suggest that the climate security agenda should seek to follow the tough but successful path that the R2P agenda took from idea phase through to institutionalisation at the highest levels of the UN, especially the UN Security Council.[39] It is also worth mentioning that the 'responsibility to protect' phrase (but unrelated to R2P) features regularly in climate action advocacy – for instance, the argument may proceed that a particular country or political leader has a responsibility to protect 'the Pacific'[40], 'the planet',[41] 'the Earth',[42] 'children',[43] or 'future generations'[44] from climate impacts.

In terms of key definitions, the book uses the expression 'climate action' to denote 'efforts to reduce greenhouse gas emissions and strengthen resilience and adaptive capacity to climate-induced impacts'.[45] In other words, the expression covers both climate mitigation and climate adaptation policies, and, importantly, is solutions-focused. The book uses the expression 'climate change' when referring generically to the science of climate change (e.g. discussion of warming, causes, or impacts) as well as to refer to climate action policies. 'Mass atrocities' and 'atrocities' are used interchangeably to describe instances of 'genocide, war crimes, ethnic cleansing and crimes against humanity', consistent with the R2P framework.[46] Efforts have been made to differentiate mass atrocities from atrocities where appropriate (e.g. larger scale versus smaller scale instances of deadly violence).[47]

It is also important to note that while atrocities are generally committed in the context of armed conflict situations (e.g. wars and smaller scale violence), they can also occur in the absence of an armed struggle (e.g. currently in North Korea and China).[48] In addition, while the tools to prevent and respond

to atrocities and armed conflict are significantly compatible,[49] distinctions can emerge in various contexts.[50] To offer a rejoinder on these two issues, the book overwhelmingly focuses on atrocities committed in armed conflicts, and develops policy responses that consider armed conflict and atrocities as a singular entity (i.e. does not develop a separate range of options specifically for atrocity situations).[51] The book therefore on all these definitions and associated issues is consistent with Gareth Evans's 2008 book on this topic.[52]
This book is divided into two parts.

Part I presents a historical overview of the relationship between climate change and atrocities. Chapter 1 sets up the twin problems that the book seeks to address: human-induced climate change and mass atrocity crimes. The chapter examines these histories in parallel, with a few cautious links identified, commencing with pre-history times, and concluding with the present time (2022–2023). Chapter 2 builds on this discussion by outlining the major international efforts to curb global warming and prevent atrocities. This policy narrative commences in the aftermath of the Second World War and again concludes in present times (2022–2023). Chapter 3 continues this policy history by retracing the recent efforts (2010–2023) by Western governments to integrate climate change with conflict prevention and response policies and issues. The chapter then presents some initial thinking on the conceptual parallels between the climate action and R2P agendas. Part II builds on this discussion by examining the relationship between the R2P's operational framework (and policies therein) and climate action policies. To achieve this, it attempts to situate, or 'parallel' as Part I does, variety of climate action policies alongside the 46 policy options found in the R2P 'toolbox' across its prevention (Chapter 4), reaction (Chapter 5), and rebuilding (Chapter 6) phases.

Notes

1 IPCC, 2021: *Climate Change 2021: The Physical Science Basis*. Contribution of Working Group I to the Sixth Assessment Report of the Intergovernmental Panel on Climate Change (Cambridge University Press, Cambridge, United Kingdom and New York, USA, 9 August 2021). AR6 WGI is the first of three major reports that contributed to the IPCC Sixth Assessment cycle. The Working Group II contribution to the Sixth Assessment Report, *Climate Change 2022: Impacts, Adaptation and Vulnerability* was released on 28 February 2022. The Working Group III contribution, *Climate Change 2022: Mitigation of Climate Change* was released on 4 April 2022.

2 See, for instance, Kate Guy and Akash Ramnath, 'Balancing on a Knife's Edge: Climate Security Implications of the IPCC Findings', *The Centre for Climate and Security*, 9 August 2021; and 'Earth Will Hit 1.5°C Climate Limit Within 20 Years, says IPCC Report', *New Scientist*, 9 August 2021.

3 See, for example, Steven Pinker, *The Better Angels of Our Nature: Why Violence Has Declined* (New York: Penguin Books, 2011); and Joshua Goldstein, *Winning the War on War: The Decline of Armed Conflict Worldwide* (New York: Penguin Books, 2011).

4 Alex J. Bellamy, 'The Responsibility to Protect at 15: A Promise Unfulfilled', *Global Centre for the Responsibility to Protect*, 21 September 2020; Alex J. Bellamy and Ivan Simonovic, 'Introduction: Towards Evidence Based Atrocity Prevention', *Journal of International Peacekeeping*, Vol 24, 2020, p 285–304 (specifically p 294). See also, 'Conflict Trends', *Centre for Civilians in Conflict* (accessed 2022).
5 *The Responsibility to Protect*, Report of the International Commission on Intervention and State Sovereignty (Ottawa: International Development Research Centre, 2001).
6 Gareth Evans, 'State Sovereignty Was a Licence to Kill', *International Crisis Group*, 22 May 2008.
7 *The Responsibility to Protect*, ICISS, p xi.
8 Alex J. Bellamy, *The Responsibility to Protect: A Defense* (New York: Oxford University Press, 2015), p 1 and 5. See also historical accounts of R2P's policy development in Thomas G. Weiss, *Humanitarian Intervention, 3rd Edition* (Cambridge: Polity, 2016), especially Chapter 4; and Ramesh Thakur, *The United Nations, Peace and Security: From Collective Security to the Responsibility to Protect, 2nd Edition* (Cambridge: Cambridge University Press, 2017), p 272–325.
9 Gareth Evans, *The Responsibility to Protect After Libya and Syria*, Address, Annual Castan Centre for Human Rights Law Conference, Melbourne, 20 July 2012.
10 See, for instance, 'Fact Sheet: Prioritizing Climate in Foreign Policy and National Security', Statements and Releases, *The White House*, 21 October 2021. 'Global Britain in a Competitive Age: The Integrated Review of Security, Defence, Development and Foreign Policy', *UK Government*, March 2021. Climate Change and Defence Roadmap, *European External Action Service*, 9 November 2020.
11 'Concept for an Integrated Approach on Climate Change and Security', EEAS(2021)770, *European External Action Service*, 16 September 2021. See also, 'Report on the Impact of Climate Change on Migration', *The White House*, October 2021.
12 'United States Strategy to Anticipate, Prevent, and Respond to Atrocities', *US State Department*, 15 July 2022; 'From Srebrenica to a Safer Tomorrow: Preventing Future Mass Atrocities Around the World', House of Commons, 11 October 2022, p 10 and 25, and 54 (see also, witnesses' contributions from Global Centre for R2P Director, Savita Pawnday, and The Lord Ahmad of Wimbledon, and submissions number 11, 20, 21, and 26). See also, Willian Worley, 'UK Parliament Report Warns Climate Change Likely to Worsen Atrocities', *Devex*, 17 October 2022.
13 The climate action agenda's prioritisation of human protection is evidenced throughout this book via the numerous adaptation policies and programmes designed specifically to protect vulnerable populations from climate impacts; the 'ultimate objective' of the international climate regime – the United Nations Framework Convention on Climate Change – which is to avoid warming 'dangerous' to humans, and the associated emissions reduction policies to achieve that objective; the numerous statements by political leaders about the need to protect people from climate impacts and to reduce emissions; and the more than ten years of anecdotal evidence collected by the author. Clearly, the R2P agenda is about human protection.
14 Lloyd Axworthy, 'Resetting the Narrative on Peace and Security: R2P in the Next Ten Years', in Alex J. Bellamy and Tim Dunne (Eds.), *The Oxford Handbook of the Responsibility to Protect* (New York: Oxford University Press, 2016), p 980. Axworthy mentions the IPCC on p 969 as well, which is not in the index of the Handbook.

15 Lloyd Axworthy and Allan Rock, 'R2P: A New and Unfinished Agenda', *Global Responsibility to Protect*, Vol 1, 2009, p 54–69.

16 See, for example, Gareth Evans, 'Conflict Potential in a World of Climate Change', Address, 29 August 2008; Simon Adams, 'From Global Warming to Genocide Warning: Climate Change and Mass Atrocities', *Global Centre for R2P*, 4 December 2016; 'Atrocity Alert No. 173: Climate Change and the Responsibility to Protect', *Global Center for the R2P*, 25 September 2019; Neal Simpson, Nikki Marczak, Alex Bellamy, Arna Chancellor, Emily Dart, Marina Riley, Shangyue Lin, Mehar, and Kate Sisley, 'Climate Change and Atrocity Crimes: The Challenge in The Pacific', *The Asia Pacific Centre for the Responsibility to Protect,* November 2020; Simon Adams, *Mass Atrocities, the Responsibility to Protect and the Future of Human Rights: 'If Not Now, When?'* (Oxon: Routledge, 2021), see chapter titled 'Climate Change and Mass Atrocities'.

17 Sherri Goodman and Pauline Baudu, 'Climate Change as a "Threat Multiplier": History, Uses and Future of the Concept', *Centre for Climate and Security*, 3 January 2023.

18 Alex J. Bellamy, 'The Responsibility to Protect: A Wide or Narrow Conception?', in Peter Hilpold (Eds.), *The Responsibility to Protect: A New Paradigm of International Law?* (Leiden: Brill Nijhoff, 2015), p 38–59.

19 See, for instance, Gareth Evans, 'The Responsibility to Protect in Environmental Emergencies' and Edward C. Luck, 'Environmental Emergencies and the Responsibility to Protect: A Bridge Too Far?', in Proceedings of the Annual Meeting (American Society of International Law), Vol. 103, International Law as Law, 2009, p 27–32 and 32–38. See also, Alex J. Bellamy, *The Responsibility to Protect: A Defense*, p 13.

20 Roberta Cohen, 'The Burma Cyclone and the Responsibility to Protect', *Global Responsibility to Protect*, No 1, 2009, p 253–257; Rebecca Barber, 'The Responsibility to Protect the Survivors of Natural Disaster: Cyclone Nargis, a Case Study', *Journal of Conflict & Security Law*, Vol 14, No 1, 2009, p 3–34; Jurgen Haacke, 'Myanmar and the Responsibility to Protect, and the Need for Practical Assistance', *Global Responsibility to Protect*, No 1, 2009, p 156–184; Mely Caballero-Anthony and Belinda Chng, 'Cyclones and Humanitarian Crises: Pushing the Limits of R2P in Southeast Asia', *Global Responsibility to Protect*, No 1.2 (2009), p 135–155; Alpaslan Özerdem, 'The "Responsibility to Protect" in Natural Disasters: Another Excuse for Interventionism? Nargis Cyclone, Myanmar', *Conflict, Security & Development*, Vol 10, No 5, 2010, p 693–713; Ashley McLachlan-Bent and John Langmore, 'A Crime Against Humanity?: Implications and Prospects of the Responsibility to Protect in the Wake of Cyclone Nargis', *Global Responsibility to Protect*, No 3, 2011, p 37–60; Julian Junk, 'Testing Boundaries: Cyclone Nargis in Myanmar and the Scope of R2P', *Global Society*, Vol 30, No 1, 2016, p 78–93.

21 Gareth Evans, 'Conflict Potential in a World of Climate Change'; Jess Gifkins, 'Darfur', in Alex J. Bellamy and Tim Dunne (Eds.), *The Oxford Handbook of the Responsibility to Protect* (New York: Oxford University Press, 2016), p 718. Many authors address climate-related impacts and R2P in the Darfur conflict, but are not R2P specialists; see, for instance, Andrew S. Natsios, *Sudan, South Sudan and Darfur: What Everyone Needs to Know* (New York: Oxford University Press, 2012); Mahmood Mamdani, *Saviours and Survivors: Darfur, Politics, and The War on Terror* (New York: Pantheon Books, 2009); Julie Flint and Alex de Waal, *Darfur: A New History of a Long War* (New York: Zed Books, 2008); Alex de Waal, 'Sudan: The

Turbulent State', in Alex de Waal (Ed.), *War in Darfur: and The Search for Peace* (Harvard University, 2007), p 28–29. Another large body of scholarship addresses the role climate-related factors played in the Darfur situation, but do not address R2P; see, for instance, Jeffrey Mazo, 'Darfur: The First Modern Climate-Change Conflict', *The Adelphi Papers*, Vol 49, 2009, p 73–86; John Hagan and Joshua Kaiser, 'The Displaced and Dispossessed of Darfur: Explaining the Sources of a Continuing State-led Genocide', *The British Journal of Sociology*, Vol 62, No 1, 2011, p 1–24; Lyal S. Sunga, 'Does Climate Change Kill People in Darfur?', *Journal of Human Rights and the Environment*, Vol 2, No 1, 2011, p 64–85; Harry Verhoeven, 'Climate Change, Conflict and Development in Sudan: Global Neo-Malthusian Narratives and Local Power Struggles', *Development and Change*, Vol 42, No 3, 2011, p 679–707.

22 Syria is major research theme in the R2P literature; see, for example, 'The Responsibility to Protect Bibliography' on the European Centre for the Responsibility to Protect. On the climate side, fierce debate has arisen over the role drought played in driving Syria's civil war; see, for instance, Colin P. Kelley, Shahrzad Mohtadib, Mark A. Cane, Richard Seager, and Yochanan Kushnir, 'Climate Change in the Fertile Crescent and Implications of the Recent Syrian Drought', *PNAS*, Vol 112, No 11, March 2015, p 3241–3246; Jan Selby, Omar S. Dahi, Christiane Frohlich, and Mike Hulme, 'Climate Change and the Syrian Civil War Revisited', *Political Geography*, Vol 60, 2017, p 232–244; Colin P. Kelley, Shahrzad Mohtadi, Mark Cane, Richard Seager, and Yochanan Kushnir 'Commentary on the Syria Case: Climate as a Contributing Factor', *Political Geography*, Vol 60, September 2017, p 245–247. See also, Marwa Daoudy, *The Origins of the Syrian Conflict: Climate Change and Human Security* (Cambridge: Cambridge University Press, 2020).

23 Joanna Harrington, 'R2P and Natural Disasters', in Andy Knight and Frazer Egerton (Eds.), *The Routledge Handbook of the Responsibility to Protect* (Routledge: Abingdon, 2012), p. 141–151. The term 'natural disaster' is used in the R2P literature much more frequently compared to the term 'climate change'. The trouble is that the former can and does in many instances encompass non-climate-related human protection situations such as earthquakes or oil spills – see, for example, Tyra Ruth Saecho, 'Natural Disasters and the Responsibility to Protect: From Chaos to Clarity', *Brooklyn Journal of International Law*, Vol 32, No 2, 2007, p 663–708. It should also be noted that a subterranean theme in the R2P literature examines the links between climate impacts, people movement and R2P – see, for example, Sara Davies, 'A Responsibility to Protect Persons in the Event of Natural Disasters', in S. E. Davies and L. Glanville (Eds.), *Protecting the Displaced: Deepening the Responsibility to Protect* (Netherlands: Martinus Nijhoff Publishers, 2010) p 163–184.

24 Gareth Evans, 'Facing Up to Our Responsibilities', Opinion Editorial, *The Guardian*, 12 May 2008; Llody Axworthy, 'It's Time to Intervene', Opinion Editorial, *Ottawa Citizen*, 13 May 2008; Gareth Evans, 'Time for an Aid Invasion?', Opinion Editorial, *The Age*, 19 May 2008; 'Cyclone Nargis and the Responsibility to Protect', Myanmar/Burma Briefing No 2, *Asia-Pacific Centre for the Responsibility to Protect*, 16 May 2008.

25 See, for example, Ken Conca, Joe Thwaites, and Goueun Lee, 'Climate Change and the UN Security Council: Bully Pulpit or Bull in a China Shop?', *Global Environmental Politics*, Vol 17, No 2, May 2017, p 13–15.

26 Linda A. Malone, 'Green Helmets: Eco-Intervention in the Twenty-First Century Unilateral and Multilateral Intervention', Proceedings of the Annual Meeting (American Society of International Law), Vol 103, 2009, p 19–27.
27 Robyn Eckersley, 'Ecological Intervention: Prospects and Limits', *Ethics and International Affairs*, Vol 21, No 3, p 293–316.
28 Charlotte Ku, 'The UN Security Council's Role in Developing a Responsibility to Respond to the Climate Change Challenge', in Shirley V. Scott and Charlotte Ku (Eds.), *Climate Change and the UN Security Council* (Cheltenham: Edward Elgar, 2018) p 162–184.
29 Ken Conca, Joe Thwaites, and Goueun Lee, 'Climate Change and the UN Security Council?', p 14.
30 Ibid., p 14.
31 Both Human Security and Environmental Peacebuilding originated in the immediate post-Cold War period as novel approaches to security, and have since spawned voluminous literatures.
32 See, for example, Saleem H. Ali and Rebecca Pincus, 'The Role of the Military in Environmental Peacebuilding', in Ashok Swain and Joakim Ojendal (Eds.), *Routledge Handbook of Environmental Conflict and Peacebuilding* (London: Routledge, 2018) p 306–317.
33 For an analysis of negative side effects of EP on peace, the environment, and development, see Tobias Ide, 'The Dark Side of Environmental Peacebuilding', *World Development*, Vol 127, 2020, p 1–9.
34 Tobias Ide, Carl Bruch, Alexander Carius, Ken Conca, Geoffrey D. Dabelko, Richard Matthew, and Erika Weinthal, 'The Past and Future (s) of Environmental Peacebuilding', *International Affairs*, Vol 97, No 1, 2021, p 2–3; Florian Krampe, Farah Hegazi, and Stacy D. Van Deveer, 'Sustaining Peace Through Better Resource Governance: Three Potential Mechanisms for Environmental Peacebuilding', *World Development*, Vol 144, 2021, p 2.
35 Richard A. Matthew and Oli Brown, 'From Conflict to Peacebuilding: The Role of Natural Resources and the Environment', UN Environment Program, 2009. See also, Tobias Ide, McKenzie F. Johnson, Jon Barnett, Florian Krampe, Philippe Le Billon, Lucile Maertens, Nina von Uexkull & Irene Vélez-Torres, 'The Future of Environmental Peace and Conflict Research', *Environmental Politics*, 2023, p 11; Ide et al., 'The Past and Future (s) of Environmental Peacebuilding', p 5.
36 Florian Krampe, 'Toward Sustainable Peace: A New Research Agenda for Post-Conflict Natural Resource Management', *Global Environmental Politics*, Vol 17, No 4, November 2017, p 3. See also, McKenzie F. Johnson, Luz A. Rodríguez, and Manuela Quijano Hoyos, 'Intrastate Environmental Peacebuilding: A Review of the Literature', *World Development*, Vol 137, 2021, p 2.
37 Ibid.
38 Shiloh Fetzek and Louise van Schaik, 'Europe's Responsibility to Prepare: Managing Climate Security Risks in a Changing World', *The Centre for Climate and Security*, June 2018; Caitlin Werrell and Francesco Femia, 'The Responsibility to Prepare and Prevent: A Climate Security Governance Framework for the 21st Century', *The Centre for Climate and Security*, October 2019.
39 See also Matt McDonald, *Ecological Security: Climate Change and the Construction of Security* (Cambridge: Cambridge University Press, 2021), p 155.
40 'Australia Must Help Protect Pacific from Climate Change, PNG Prime Minister says', *The Guardian*, 27 July 2019.

41 António Guterres, 'Remarks at 2019 Climate Action Summit', September 2019.
42 Stewart M. Patrick, 'A Responsibility to Protect the Earth? Reframing Sovereignty in the Anthropocene', *World Politics Review*, 3 February 2020.
43 'How to Talk to Children About Climate Change', *ABC News*, 18 December 2019.
44 Jeff Tollefson, 'Kids Climate Lawsuit to go Before Alaska Court', *Nature*, 2019.
45 'Climate Action', UN Development Programme (accessed 2021).
46 Gareth Evans, *The Responsibility to Protect: Ending Mass Atrocity Crimes Once and For All* (Washington, DC: Brookings Institution Press, 2008), p 11–13.
47 David Scheffer argues that 'for all policy discussions purposes one should just use the generic expression "atrocity crimes"' and leave it to the courts to sort out the other details on a case-by-case basis – see Evans, *The Responsibility to Protect* (Brookings, 2008), p 13.
48 Alex J. Bellamy, 'Mass Atrocities and Armed Conflict: Links, Distinctions, and Implications for the Responsibility to Prevent', *The Stanley Foundation*, February 2011, p 1; 'From Srebrenica to a Safer Tomorrow: Preventing Future Mass Atrocities Around the World: Government Response to the Committee's Third Report (Appendix: Government Response)', House of Commons, UK Government, 17 January 2023, p 1 and 3.
49 Bellamy, 'Mass Atrocities and Armed Conflict', p 1 and 6.
50 Ibid., p 4–5.
51 This approach is supported in Bellamy, 'Mass Atrocities and Armed Conflict', p 6. See also the establishment of the UK Mass Atrocity Prevention Hub in 'From Srebrenica to a Safer Tomorrow', UK Government, 2023, p 1 and 3.
52 Evans, *The Responsibility to Protect* (Brookings, 2008).

Part I

Climate Change and Atrocities

The History

1 A History of Climatic Change and Atrocities

> Earth's climate has changed throughout history. Just in the last 650,000 years there have been seven cycles of glacial advance and retreat, with the abrupt end of the last ice age about 11,700 years ago marking the beginning of the modern climate era – and of human civilization.
>
> (Global Climate Change: Vital Signs of the Planet, *NASA*, 2022)

> Massacres of the innocent, forced displacement of populations, large-scale sexual violence and humiliation, and the wanton destruction of civilian property have been going on since the dawn of civilization.
>
> (Gareth Evans, *The Responsibility to Protect: Ending Mass Atrocity Crimes Once and For All*, p 11)

This chapter presents an overview of the parallel problems of climatic change and mass atrocities. The ultimate aim is to provide a clear sense of the two problems individually by way of presenting two historical narratives. However, the chapter also suggests that a complex relationship exists involving climate change, between the rise and fall of economic power and atrocity situations. The chapter begins by examining natural climate variation and atrocity situations during the pre-industrial period (pre-1760). It shows that in this period, changing weather patterns determined the subsistence livelihoods, and survivability, of much of the world's population and that mass atrocities were committed with indifference, if not applause. This section also suggests that the fate of ancient and Medieval political and economic systems, and the populations therein, were somewhat determined by dramatic changes in the climate system. The second section examines the period from the Industrial Revolution to the end of the Second World War (1760–1945). It shows that coal and later oil were central to the West's economic rise (and to those countries beyond), which repeatedly manifested in imperialist polices that resulted in countless mass atrocities the world over. The third section examines the Cold War period though to the turn of the new millennium (1950–2000). It shows that mass atrocities were the upshot of decolonisation combined with an intense ideological struggle over competing economic systems. The fourth section covering the period from the September 2001 attacks to the Covid-19

DOI: 10.4324/9780367201364-3

pandemic (2001–2022), shows that fossil fuels remain central to the distribution of economic power (and by implication, military power) in world politics, and thus continues to play an indirect role in policies that have resulted in mass atrocities.

The chapter also shows that from 1760 to 2022–2023, the concentration of carbon dioxide in the atmosphere increased from 280 parts per million (ppm) to 419 ppm and the average global temperature increased by roughly 1.1 degrees Celsius (°C).

These findings are the result of initially reading separate literatures on the history of industrialisation, fossil fuels, global warming, atrocities, and war, then subsequently in combination, rather than seeking out historical information that confirmed my pre-existing view. The chapter does not delve into the theoretical or practical complexities of the climate-conflict relationship (these relationships are examined throughout the book in the context of the Responsibility to Protect (R2P). Rather, the chapter simply suggests that while a multitude of direct and indirect foreground relationships have been examined (e.g. the 'resource wars' thesis verses the 'threat multiplier' thesis, respectively),[1] a potential novel way to understand the climate-conflict relationship is through a deep historical analysis that identifies the presence of an intertwined background relationship between climate change, economic power, and conflict (and by implication, atrocities).[2]

The chapter begins with the pre-industrial period.

Pre-Industrial Period (Pre-1760)

Mass atrocities are inseparable from the human record.[3] As archaeological records show, Palaeolithic Stone Age societies were well accustomed to indiscriminate violence, with small-scale massacres (below 50 people) a core part of life and survival.[4] The transition from small and dispersed hunter-gather populations to large and centralised pastoralist populations, some 12,000 years ago, increased the scale of the human carnage.[5] To help protect their now sedentary populations and economic assets, Neolithic societies invented new military technologies such as the dagger and most importantly the bow and arrow.[6] Global average temperatures transitioned from cool to warmer and back to cool, in what researchers term the 'interglacial period', which had a variety of impacts.[7] For example, recent research has found that the Neanderthals disappeared as the result of human-driven extermination and climatic change.[8]

Most historical accounts of atrocities commence however with the growing economic and military might of Athens. The standout case study in this period examines the Athenian destruction of the island of Melos during the Peloponnesian War (400 BC), which involved the slaughter of 'all the men of military age' and the enslavement of the women and children, according to Thucydides.[9] Elsewhere in the pre-modern world, kings and generals, such as the Assyrians (900 BC to 600 BC), who ruled across the present-day Middle East and North Africa, would routinely commit large-scale atrocities as a demonstration of military success, empire validation, and prestige.[10] Another major case study in the atrocity literature is the Roman Siege and eventual razing of Carthage during the Third Punic War (149 to 146 BC) that saw at least 150,000 Carthaginians perish out of a total population of 200,000–400,000.[11] During Europe's Medieval period (400s to 1500s AD) mass atrocities would continue to be committed with indifference, including, for example, Catholics would commit mass murder against 'non-believers' in various Crusades in France (against Cathar heretics), in Germany (against the Jewish population), and across the Middle East.[12]

Researchers have found that major changes in weather patterns in pre-modern times, in particular instances of prolonged severe drought, have coincided with the collapse of previously powerful civilisations, including the Anasazi, Akkadian (Mesopotamia), Mayan (Guatemala), Mochica (north Andes), and Tiwanaku (Bolivia) empires.[13] However, fierce debate continues over the degree to which climatic events caused the downfall of these early human civilisations. Some argue that over the last millennium, and earlier, 'megadroughts forced societal collapse: they presented insurmountable obstacles in rainfed-agriculture regions and reduced river flow conditions in irrigation-agriculture regions, thereby undermining the economy and political structures of agriculture dependent societies'.[14] Others however, rejecting 'climate determinism' – the notion that climatic change alone caused the collapse of some early civilisations – argue that a multicausality approach provides a more accurate explanation of societal collapse, which involves the analysis of the complex interplay and tensions between existing political, social, and economic factors that have been amplified as a consequence of a major climate event such as a protracted drought.[15] The evidence is mixed as to whether the political and societal collapse of major ancient civilisations – that corresponded to instances of significant climatic change – produced war and, by implication, mass atrocities. The collapse of Classical Maya is one of the most prominent cases in the affirmative (the collapse of Angkor is one of the most prominent in the negative).[16]

A large variety natural data sets such as tree-rings and lake sediment cores as well as artworks reveal that Classical Maya suffered a succession of severe droughts in the Mayan Lowlands between 750 and 1100 AD.[17] An additional suite of evidence collected by researchers show that this period of time corresponded to the gradual sociopolitical fragmentation and decline of the civilisation, as well as the proliferation of 'inter-polity warfare' and increasingly frequent mass atrocity events.[18] The evidence of atrocities include depictions in artwork on Classical monuments that show weaponry and sacrificial scenes and inscriptions such as 'to chop', which suggests executions after warfighting events;[19] the widening distribution of military fortifications such as defensive walls, and weapons such as the scattering of spear and dart points;[20] and the scale and distribution of mass graves and 'skull pits' that include men, women and children.[21] This body of research tends to indicates that attacks often targeted residential zones, with homes showing evidence of being rapidly abandoned and then burnt down.[22] Evidence has also been assembled that finds a correlation between Maya's drought period and a dramatic increase of large-scale sacrifices that some suggest were to appease the 'weather Gods'. This finding is consistent with similar regional cases, such as the 227 meticulously arranged children discovered in an ancient mass grave in modern-day Peru that archologists explain 'were sacrificed to appease the El Nino phenomenon' – in other words, to avoid a major flooding event destroying their agricultural system.[23] Ultimately, the causes of violent conflict, and atrocities, in the Mayan case (like the Peru example) have been identified as drought-induced food shortages and inflation that undermined the legitimacy of the ruling class vis-à-vis the non-elites, and or the increased competition over dwindling natural resources between regional rulers.[24]

The Mayan collapse occurred during what climate scientists and historians refer to as the Medieval Warm Period (MEW), spanning roughly 800 to 1200 AD. During these centuries, planetary temperatures increased, some suggest, by roughly 1°C.[25] Scientists explain that this period of pre-industrial warming was likely due to solar radiation, among other natural factors, and in no way discredits the fact that the build-up of carbon dioxide in the atmosphere since the Industrial Revolution is directly linked to the parallel increase in global land and sea surface temperatures. The MEW period also corresponded to the rise of Genghis Khan's army who laid waste to much of Eurasia and China. Researchers contend that under Khan's direction 'entire nations were exterminated, leaving behind nothing but rubble, fallow fields and bones'.[26] Others write that 'if genocide did not take place, conquest seemed incomplete' for the Mongol army.[27] Khan's army is estimated to have killed 40 million people between 1206 and 1227.[28] Khan's success in conquest, however brutal, was a consequence of his brilliant leadership qualities, skills as a military strategist and administrator, but also, some historians argue, due to 'the realities of the Medieval climate' that allowed, and in some instances forced, the movement of his nomadic genocidaires.[29]

While all this was occurring in modern-day Asia and the Americas, farmers in the latter centuries of Medieval Europe (1100–1300) were basking in slightly warmer and drier conditions, but with good rainfall.[30] The effects of improved climatic conditions manifested in all kinds of subtle ways – in better harvests and fishing, population growth, and increased construction projects and trade.[31] Vineyards flourished in England.[32] But in a paradox to the Mayan experience, evidence suggests that climate-forced abundance (compared to scarcity) also brought increased levels of violence. Increasingly abundant food supply meant that armies could be better fed and equipped and long-harboured strategic goals finally actioned – 'warfare was endemic in the warm centuries' of Medieval Europe, researchers explain.[33]

The MEW period gave way to the Little Ice Age (LIA) – a period from 1350 to 1800.[34] By the 1400s, the weather had become cooler, stormier, and more unpredictable compared to the MEW, with sudden climate extremes such as frosts and snow and generally lower temperatures that culminated in the bitter cold decades of the late 1500s and 1600s – depicted in a 1687 painting of a frozen Thames in London. In the Northern Hemisphere as a whole, 1641 was the third-coldest summer recorded over the past six centuries, 1643 was the tenth-coldest, and 1642 was the twenty-eighth coldest: three landmark winters in a row.[35] Climatologists have found that the LIA coincided with the lowest solar energy received on earth in two millennia, perhaps due to fewer sunspots and or a spate of major volcanic eruptions between 1635 and 1645 that deflected the Sun's radiations, combined with the increase in the occurrence of the El Nino effect in the mid-1600s. The LIA period corresponded to a huge uptick in political upheaval and warfare.[36] The two most prominent case studies are China and Europe.[37]

As an overwhelmingly agrarian society, the livelihoods of the Chinese peasantry were intertwined with climatic conditions. By examining meticulously kept historical archives on major events, military actions, and the economy, researchers have discovered that the LIA as experienced in China corresponded with greater food insecurity and famine amongst the Chinese peasanty and an increased number of rebellions and warfighting, costing millions of lives.[38] The outbreak of these LIA wars, other research adds, eventually caused dynastic collapse and a transition of power from Ming to Qing in the mid-1600s.[39] Some researchers estimate that the fall of the Ming Dynasty resulted in 25 million deaths.[40] Similar circumstances would befall Europe during the LIA. Reconstruction of temperature records combined with a 1000-year history of violent conflict in Europe reveals that the 'cold weather coincided with violence'.[41] Indeed, historical records show the 1500s and 1600s in Europe to be 'the most warlike', with a war underway in 95 per cent of the years.[42] As could be expected, human mortality was high. For example, estimates of military and civilian deaths as a consequence of the Thirty Years' War (1618 to 1648), a conflict primarily fought in Central Europe, range from 4.5 million to 8 million – the majority dying from disease or starvation, but violent deaths and mass atrocities commonplace. The Peace of Westphalia, negotiated in 1648, would not stop subsequent civil wars in France, referred to as the Fronde Revolt (1648–1653), which killed thousands more. This dire situation in Europe led Thomas Hobbs to famously write in 1651 that 'the life of man (is), solitary, poor, nasty, brutish, and short'.[43]

The extent to which the cooler temperatures of the LIA contributed to the increased occurrence of violent conflict in Europe is a matter of conjecture. On the one hand, again, similarly to the Chinese situation, some researchers show that Europe, at the time, was a rural continent with farmers and their families still living from harvest to harvest, and therefore the agro-economic downturn as a consequence of the cooler weather was indeed a 'direct cause' of Europe's large-scale pre-industrial human crises.[44] On the other hand, some authors caution not to 'paint bull's-eyes around bullet holes'.[45] Views on this side of the debate range from the cooler temperatures 'fatally exacerbating' the major political upheavals of the time[46] through to others who remonstrate that 'it would be quite absurd' to explain the Fronde by the adverse meteorological conditions of the 1640s.[47] But consensus can be found in the proposition that Medieval humans, similarly to their ancient ancestors, tended to blame supernatural forces (God's punishment for wrongdoing, or the devil's work on earth carried out by witches) for the erratic climatic conditions.[48]

The LIA also corresponded with an upsurge of industrial activity in Europe and voyages of conquest and settlement. Medieval Europe had traditionally used wood to provide for much of its construction and heating (home and industrial) needs. In the mid-1500s however, due to concern about wood shortages, Britain began to transition from wood to coal as the main source of heating. The late 1500s would also see the Spanish invasion, occupation, and exploitation of Central and South America with what some scholars describe as, 'the most extensive and most violent program of human eradication that the world has ever seen'.[49] Across present-day Mexico, Peru, Bolivia, and Ecuador, the conquistadors would massacre tens of thousands of people in coordinated village-to-village campaigns on the Aztec and Inca empires, with those not murdered worked to death in gold mines.[50] In 1608, British Pilgrims arrived in Massachusetts and commenced a settler expansion across the present-day United States (US) that wreaked havoc on the Indian population. Disease was among the biggest killer; however, genocidal massacres of men, women, and children were also prominent – for example, in the Pequot War (1636–1637) – reducing the Indian population from 7–10 million to about 230,000 between the 1600s and 1800s.[51] Other methods of displacement involved the 'deliberate destruction of flora and fauna that the American Indians used for food and other purposes', for example, the destruction of buffalo – considered an 'ecocide campaign' – that destroyed the core of their spiritual and ceremonial world.[52]

Back in Britain, concern was growing in government ranks that the increasing rate of deforestation to service the growing needs of the transport (ships and wagons), agricultural (farms and pastures), and heavy industry (mines and foundries) sectors would eventually imperil the British navy. More coal (mining and usage), especially for heating, was identified as the solution. Metallurgy and mining gained ground rapidly with coal replacing wood in the heating process for converting wrought iron into steel between 1612 and 1620.[53] Coal-burning soon became cheaper than wood-burning, and in the late 1600s and 1700s, one industry after another converted to coal including nonferrous smelting (lead and tin ores), coppers, glass making, brickmaking (with mortar in the walls made from coal-burned limestone), saltpetre, gunpowder, starch, candles, preserved foods, vinegar, whisky, beer, sugar, soap, and finally in the 1780s pig and bar iron.[54] The adoption of coal changed the economic history of Britain, then the rest of Europe, and finally the world, but also produced increasingly large quantities of carbon dioxide.

Modern climate science reveals that for the millennium before 1800 ('pre-industrial times'), the concentration of carbon dioxide in the earth's atmosphere was rather stable at roughly 280 ppm.[55] Preeminent climate scientist John Houghton explains that 'for several thousands of years before the beginning of the Industrial Revolution around 1750, carbon dioxide as measured from ice cores kept within about 10 ppm of a mean value of about 280 ppm'.[56] Temperature variations of roughly 1°C (plus and minus) that occurred in the MEW period and subsequent LIA therefore cannot be explained by increases in greenhouse gases, especially carbon dioxide or methane. Rather, Houghton continues, variations in volcanic activity combined with variations in the Sun's energy output can explain these periods of climate variability.[57] But all this would change with the Industrial Revolution.

The Industrial Revolution to the Second World War

Historians explain that stripped to its bare bones 'the industrial revolution consisted of the application of new sources of power' to production processes.[58] Despite early coal-fired advances, by 1700, Britain still overwhelmingly relied on wood, human and animal power, with some small assistance from waterwheels, for almost all of its production processes in key sectors of the economy, namely, heavy industry (manufacturing, metallurgy, and mining), transport, and agriculture.[59] However, during the 1700s, the primary sources of power transitioned from these rudimentary options to increasingly sophisticated motors powered by fossil fuels, initially coal, then later petroleum (producing incrementally larger amounts of carbon dioxide).

In the 1700s, British manufacturers learned how to produce coke from coal, and used coke instead of wood-derived charcoal for smelting iron ore. As coke production grew so too did blast furnace size and technology. By the late 1700s, coal-fired steam powered machines for rolling metal began to replace manual hammer and anvil technology. The British iron industry expanded rapidly. In 1720, Britain had produced 25,000 tonnes of unrefined iron ('pig iron'); by 1796, the figure was 125,000, and by 1804, the figure was 250,000.[60] The growth of the iron industry caused a surge in coal mining to provide fuel for iron smelting and for steam engines generally.[61] Initially, wood and metal rails were laid down to access increasingly deeper coal faces to increase production. However, such was the demand that in 1804, the first steam engine for hauling coal out of mines on tracks was introduced, which would gradually replace horses and people power. In 1830, Britain produced about 24 million tonnes of coal, four-fifths of the world's total; by 1870, the figure was 110 million tonnes – still half of all coal mined around the world.[62] By 1830, British pig iron production had grown to 700,000 tonnes per year; by 1860, production had more than quadrupled to almost 4 million tonnes.[63]

Coal-fired power combined with innovative machines also revolutionised the British textile industry. The major advance came in the 1760s with the 'flying shuttle' – an automatic weaving machine that could replace human power. Soon after, coal-fired steam engines began to power cotton spinning machines. By 1785, the 'power loom' had been developed that fully integrated the previous technologies with steam engines. Steam-powered textile factories begin to proliferate. By 1850, there were 260,000 power looms operating across

Britain, with cotton production providing about half of the entire value of British exports. As industrial outputs grew, Britain's horse and wagon powered transportation networks came under increased pressure. In 1821, a coal powered steam locomotive was trialled in northern England on a line built between a coal mine and seaport. By 1830, there was about 200 kilometres of railway line in Britain; by 1871, this figure had jumped to more than 20,000 kilometres. By the late 1800s, locomotives were omnipresent in Britain and the symbol of the Industrial Revolution.[64] In 1838, the first transatlantic steam powered shipping line was opened.

One of the deeply disturbing aspects of the West's industrialisation was the use of slave labour. The peak of the transatlantic slave trade was reached during the 1780s, with more than 75,000 enslaved people brought to the Americas each year. About half the captives were transported from Africa in ships of British merchants.[65] Britain's cotton was mostly imported from slave plantations. The transatlantic slave trade between 1452 and 1807 (after which it was abolished in Britain) resulted in 16 million deaths, with mass atrocities a matter of course – indeed some scholars have classified the trade as genocidal based on the 'incalculable demographic and social losses on West African societies, utilising every genocidal strategy listed in the UN Genocide Convention definition'.[66]

The coal-fired Industrial Revolution, which first took shape in Britain in the late 1700s, would soon expand to Western Europe: Belgium, Germany, France, and to the new United States.[67] Belgium was the first country in continental Europe to experience industrial transformation.[68] Over the 1800s, Belgium would build an extensive rail network, factories, and mining operations. Rich coal reserves and proximity to Germany and France (compared to Britain) presented an opportunity to develop an export-orientated economy in manufactured goods and raw commodities. Exports initially included pig iron and large volumes of coal to France; and later, steel and metallic products for steam engines, locomotives, and weapons. Belgium's early mover status would make it the foremost industrial power in continental Europe by the mid-1800s.[69]

Germany's industrialisation commenced after 1830, with policy and governance shifts that permitted the freedom of labour and material, and the rapid expansion of coalmining. Between 1840 and 1870, Germany's coal production expanded sevenfold, with iron production expanding 14 per cent per year in the 1850s. Germany's railway network also expanded massively during this period. In 1870, German industry benefited from the acquisition of Alsace and the minerals rich Lorraine region after defeating France in the Franco-Prussian war (1870–1871) – a conflict that cost hundreds of thousands of military and civilian lives. By 1913, Lorraine alone produced 47 per cent of all iron ore mined in Europe. While starting late, Germany's excellent coal and iron resources delivered rapid industrialisation from the 1850s onward.

Industrialisation in France, Western Europe's richest country in the 1700s, was severely hampered by recurrent revolutions in 1789, 1830 and 1848. In 1789, the most notable of the three, French rebels overthrew King Louis XVI (who was subsequently executed in 1793), which triggered a civil war – the French Revolution. While mass atrocities were committed by both sides, a particularly egregious situation, which some scholars have called a genocide, took place across the Vendee region where up to 150,000 people, civilians, and combatants, were shot, drowned, bayonetted, hanged, or guillotined on government orders.[70] France's economic transition was also constrained by limited coal and mineral reserves (owing to Germany). Having to import expensive coal (from Belgium), and war ravaged, caused France's relative economic position to decline in the 1800s. However, industrialisation would proceed. French inventions to advance mechanisation in the textile industry, combined with several new metallurgical hubs that massively expanded iron production, and a government built national railway network, helped to drive a thirteenfold increase in France's coal production between 1820 and 1870.[71]

Having won its independence from Britain in the American Revolutionary War (1775–1783), industrialisation exploded in the US between 1850 and 1900. The US began to import new technology from Britain in the 1820s, and later Germany and France, which was quickly imitated and improved on.[72] By 1850, a vast network of railway tracks had been laid down and populated with locomotives for transport and manufacturing. Coal mining expanded, steelmaking boomed, and the textiles industry grew. In 1861, the first US oil refinery commenced operations. The same year, the American Civil War (1861–1865) began which reorientated American manufacturing towards armaments and associated war industries. When the war ended, and the domestic market for arms sales shrank, these industries began to export arms abroad.[73] With the discovery and refining of oil, agriculture too would be revolutionised, with horse-drawn harvesting giving way to tractors operating across the US's vast agricultural planes. Mass atrocities targeting Apache, Cherokee, and Navajo would continue, including a devastating state sanctioned extermination campaign against the Yuki between 1860 and 1880, reducing the population from 20,000 to just 168.[74] Children were systematically removed from their families with a high mortality rate among those interned (a system that became notorious in Canada).[75] By 1880, the US was responsible for 85 per cent of the world's crude oil production and refining.[76]

By the late 1880s, the bulk of Western society had been transformed into coal-fired and carbon-intensive industrialised economies.[77] The US and Germany (and to a lesser extent, Belgium and France) were now spearheading technological innovation and change, moving well beyond Britain's first efforts. Economic competition and political rivalry in Europe, combined with an upsurge of social pressures and a depression, again stirred Europe's imperialist instincts.[78] To gain economic advantage, Europe's major economies would seek to establish and monopolise new markets in Africa and Asia and then insulate those markets from European competitors, as well as ruthlessly exploit the natural and human resources on offer.[79] Advances in coal-fired metallurgy and machines would be central to achieving these goals. For instance, new steam-powered ships and locomotives would be used to deploy a greater number of troops over quicker timeframes and distances; and they would carry new mass produced, and dramatically more lethal, weapons such as the 'maxim gun' (the world's first machine gun – 1884).[80]

The cornerstone of Europe's 'New Imperialist' agenda (1880–1914) was the invasion, annexation, division, and colonisation of Africa, otherwise known as the 'Scramble for Africa'.[81] Examples of mass atrocities abound. Belgium imperialism in the Congo Free State (1895–1908) and its reign of terror for control over natural resources resulted in 10 million deaths, with males worked to death, children and women routinely disfigured as punishment, and the wholesale slaughter of countless villages across the region.[82] Atrocities and disease reduced Congo's population by half by the 1920s.[83] German colonisation of Southwest Africa, present-day Namibia, in 1903 – primarily to establish and expand agricultural enterprises – involved the forced labour, detention, maltreatment, and murder of the local Herero nation. Indeed, by 1908, the occupying Germans had almost entirely exterminated the Herero people, with accounts of men hanged in rows, women raped and bayoneted, and children corralled into log enclosures and burned alive.[84] By the late 1800s, France had colonised as many as 29 countries in Africa, most located in West Africa (e.g. from Algeria, then east to Chad and back to the west coast such as Guinea). Standing behind an outward 'civilising mission' reports from Africa spoke of thousands of workers being taken from villages by force, often lassoed around the neck, and then worked to death on massive railway projects.[85] In Algeria, colonisation and atrocity conspired to reduce the native population from 3 million in 1830 to about 2 million by 1860.[86]

British imperialism in Africa encompassed areas to the East (e.g. Egypt through Sudan to Kenya) and the South (e.g. parts of South Africa through Zimbabwe to Zambia). Britain's military invasion of Egypt in 1882 to secure the strategically and commercially important Suez Cannel would set in train the 1890 colonisation and socio-political division of Sudan that would inflame local tensions and contribute to atrocities to this day; while in the south, two successive and brutal wars (1880–1881; 1899–1902) with the Dutch-speaking Boers would cost tens of thousands of military and civilian lives, including thousands of Boer women and children perishing in concentration camps. Researchers have found that Europe's use of Africa as its primary source of food supplies and raw materials to fuel its industrialisation (while the US used Central and South America) was 'often at the expense of the environment'.[87]

Russia's industrialisation lagged the West owing to a deeply held suspicion of the West's exploitative economic instincts and from losing the Crimean War (1853–1856) against Turkey (now, Türkiye), Britain, and the French, which resulted in hundreds of thousands of people killed. But from the 1890s Russian industrialisation would progress rapidly, owing much to its extensive coal and oil reserves.[88] Russian coal and iron mining accelerated between 1850 and 1900, directly inducing heavy industry output.[89] For example, in 1860, Russia only had about 1,000 kilometres of rail, by 1894 it had 30,000 kilometres, and by 1900 almost 60,000 kilometres, including a new line that traversed Siberia, opening its vast resources to industrial use.[90] By 1913, Russia had become the world's fourth largest steel producer. In 1870, oil was discovered in the Caucasus; by 1900, Russia was the second largest producer of petroleum in the world. Concurrently (post-1870s), Marxist leaders (based on Karl Marx's research) were spreading socialist doctrine to counter Western capitalism, and in 1898 formed the Social Democratic Workers Party.[91] After losing a war against Japan (1904–1905), costing tens of thousands of casualties, a general strike took hold across Russia, and unions, or 'soviets', were established in many factory centres. Industrialist employers resisted and the government banned the Marxist party and jailed its leaders. Yet, despite war and social strife, by 1914 Russia had managed to partly industrialise owing to a combination of eventual Western involvement in its domestic economy, its massive natural resource base and enthusiasm for expanding mining and manufacturing operations, and the central government's own military needs.[92] The subsequent civil war (1918–1923) – the Russian Revolution – that pitched the Bolsheviks 'reds' (communists) against the anti-Bolsheviks 'whites' (favouring capitalism) claimed an estimated 9 million lives.[93]

Meanwhile, in the West, a second industrial revolution of sorts was underway. By the turn of the nineteenth century two new engines had been developed and had begun to replace the steam engine: the electric turbine – powered by coal (or petroleum or hydro), and the internal combustion engine – powered by petroleum.[94] More than 200 oil by-products began to enter daily life including fuel for stoves and lubricants for industrial machinery and to power the internal combustion engine.[95] Between 1900 and 1914, car companies proliferated throughout Western Europe and the US, powered by mass-produced petrol engines. By 1927, the US would become the most motorised country in the world, with one motor vehicle for roughly every five people (and would lead auto ownership per capita for the next century).[96] By contrast, in Britain, France, and Germany only one in every 44 people would own a motor vehicle.[97] Construction was transformed using mechanical saws, petrol-powered lifting equipment, and new materials such as preformed concrete. The textiles industry adopted electric sewing machines. Food processing, from tractors and harvesters through to canning machines in factories, would use oil and coal-based power sources. Time-saving appliances such as washing machines and vacuum cleaners were introduced into homes and powered predominantly by coal-fired electricity. Almost all types of work could now be mechanised and powered by fossil fuels.[98] And the chemicals industry would develop a new oil-based material called 'plastic'.

The First World War (1914–1918) was the first full-scale demonstration of the sheer destructive power that industrial advances could bring to bear. A combination of new weapons technology and the ability to mass produce and distribute these technologies – namely, rapid-fire small arms, long-range artillery, tanks, aerial bombardment, submarines, and poison gas – delivered a previously unimaginable killing efficiency with tens of millions of soldiers killed on each side over a four-year period.[99] The clearest example of what industrialised war could achieve occurred on 1 July 1916 on the Somme in France where a major British offensive by 100,000 troops produced 60,000 casualties, including 20,000 killed, in one day.[100] The war also sparked a Turkish-led genocidal campaign against the Christian minorities of the Ottoman realm that resulted in the death of 1.5 million Armenians between 1915 and 1923.[101] In addition, the war demonstrated to the Allied Powers (Britain and France; with Russia as the Triple Entente; and the US) and Europe's Central Powers (Germany, Austria-Hungarian and the Ottoman's) that access to oil for fuelling supply trucks, tanks, ships, and planes would be critical in any future war. In an effort to secure supply and avoid price volatility, in the aftermath of the war, the US embarked on an extensive oil diplomacy campaign in the Middle East. In 1928, the Red Line Agreement was reached that would in effect transfer control of the bulk of Mideast oil production in ex-Ottoman territory (Türkiye, Iraq, Saudi Arabia, Syria, Jordan) to the US (and away from Britain and France) by the early 1930.[102]

In 1928, Joseph Stalin became supreme leader of the Soviet Union. Stalin's successive 'five-year plans' prioritised heavy industry, big factories, and innovative technologies such as the handheld coal mining drill replacing the pick.[103] Success was immediate. Between 1928 and 1935, coal production increased from 35 million tonnes per year to 109 million tonnes; pig iron production almost quadrupled, and steel production tripled; chemical and machine building expanded – for example, by 1936, the Soviets produced more tractors than any other nation. Over 1,500 new factories were built during the 1930s. And Soviet electrification rates jumped from fifteenth to third in the world between 1913 and 1935.[104] Researchers point out however that Stalin's 'new program of extremely – almost hysterically – rapid industrialisation' lay the seeds of 'two principal (interwoven) genocidal policies: the massive expansion of the Gulag system, or prisons, and the campaign against the peasantry, whose grain was needed to feed cities swelled by Stalin's crash industrialisation program'.[105]

By 1936, Stalin had dispatched approximately 2.4 million peasants and suspected political opponents to the Gulags. While some died in transit, owing to the appalling conditions, once there, prisoners were largely used as forced labour on export-orientated mining and timber harvest projects.[106] The 15 per cent mortality rate in the Siberian gulags has been described as a 'genocide implemented through intentional neglect and wilful disregard for subsistence needs'.[107] Stalin's destruction of the rural workforce combined with high taxation and grain seizures produced a famine across the Soviet Union between 1930 and 1933 that claimed the lives of an estimated 5.7 million people. The Soviets exported millions of tonnes of grain 'to buy foreign machinery intended to make possible the accelerated industrialisation plans' while millions starved, explain researchers.[108] Additional mass atrocities occurred between 1937 and 1938, as Stalin embarked on 'The Great Purge', so-called, that resulted in the execution of more than 680,000 suspected and real political opponents (anti-communists).[109]

The Second World War (1939–1945) is considered the 'largest and most destructive military conflict in history', costing an estimated 66 million lives.[110] In Europe, battles raged to the east and west of Germany. In the east, the Germans and the Allies (US, UK, France, among other Commonwealth and European countries in the west) showed a degree of respect for civilians and prisoners-of-war. By contrast, in the east, crimes against humanity, including genocide, became normalised.[111] Hitler's goal in the east was to wipe out the Soviet state to make way for an Aryan paradise west of the Ural Mountains.[112] To achieve this, German troops were granted licence to exterminate political enemies and inferior races and murder disarmed soldiers and defenceless civilians.[113] By 1943, the Soviets had pushed the Germans back to Poland. Soon after, the Soviets reached Germany. In retribution for German atrocities in the east, the Soviets unleashed a campaign of mass murder and mass sexual violence against German civilians.[114] By the end of war, Allied forces had 'area bombed' most large and medium-sized cities across Germany, which cost the lives of between 300,000 and 600,000 people – mostly civilians, the elderly, woman, and infants.[115] However, of all the mass atrocities discussed in the present chapter, the Jewish Holocaust is a clear historical beacon.

In 1933, Adolf Hitler became German Chancellor. His self-ascribed task was to restore German greatness in the world. Hitler, and his advisors, harboured an intense hatred of the Jewish population, which, upon Hitler's appointment, led to their immediate mass dismissal from German public institutions and exclusion from social services as well as the eventual removal of property, wealth, and citizenship. Soon after, tens of thousands of Jewish men were imprisoned in concentration camps, while others were murdered. The Nazis also committed mass murder against their political opponents especially communists, 'gypsies' (Romani people), people living with a disability, and the infirmed, among other minorities. This soon turned to genocidal mass murder. By 1943, more than 1.8 million Jewish men, women, children, and infants had been rounded up and murdered – at this stage, mostly herded into excavated ditches, shot at point blank range, and buried together.[116] To increase killing efficiency and to protect German executioners from any potential psychological damage incurred from murdering children (especially) at close range, industrialised death camps with gas chambers became the preferred method of murder. Nearly 1 million Jewish individuals and families were killed at the Auschwitz-Birkenau network of camps using cyanide gas. Nearly 2 million more died in Nazi extermination camps in Poland by gas, shootings, beatings, and starvation.[117] Between 1941 and 1945, 5 to 6 million Jewish people were systematically murdered by the Nazi regime, its allies, and surrogates in Nazi-occupied territories.[118]

Japanese imperialism in the early 1900s culminated in the launch of the Second World War in the Pacific in 1937 with the invasion of China's eastern seaboards under air attack, killing tens of thousands of civilians, and 'even more intense atrocities' committed by Japanese troops on the ground.[119] For example, the occupation of the Chinese capital, Nanjing, which has become 'a global byword for war crimes', involved Japanese forces slaughtering as many as 200,000 men, and raping tens of thousands of women and children – often murdering and mutilating their victims thereafter.[120] Over the course of the occupation, between 1937 and 1945, it is estimated that almost 2.5 million unarmed Chinese civilians were killed.[121] Subsequent invasion though Southeast Asia exacted a staggering death toll on local populations with the treatment of prisoners-of-war notorious. By the end of the war, the US air force had firebombed cities across Japan resulting in an estimated 900,000 civilian deaths; and in 1945, nuclear weapons were exploded over Hiroshima and Nagasaki causing an additional 90,000 and 70,000 deaths, respectively.[122]

As was forecast, oil was critical to winning the war in Europe and the Pacific. At the start of the war, the US was responsible for 60 per cent of world oil production and was supplying Europe with significant quantities. Germany's oil supplies largely came from captured territories to the east. As the war continued, Allied bombers stepped up their campaigns on German oil infrastructure, while German U-Boats sank US oil shipments largely unchallenged.[123] Japan relied heavily on US oil imports, and in mid-1941, the US effectively cut off supply. A few months later, Japan attacked Pearl Harbour, triggering full US military engagement in the Pacific campaign. However, by 1943 – amid fighting two major wars – concern was growing in the US that domestic oil production was diminishing, despite nation-wide fuel rationing. Saudi Arabia's oil supplies were subsequently identified as critical to US security interests and in early 1945, the US and Saudi Arabia agreed to cooperate on oil production. Soon after, the world's largest oil field was discovered in Saudi Arabia.

By 1945, awareness and scientific knowledge had grown about the potential damage that carbon dioxide may cause the atmosphere (unlike in previous periods where faith-based explanations of changing weather patterns reigned supreme).[124] In the 1820s, French scientists drew comparisons between the atmospheric layer and the glass in a greenhouse, leading to the name, 'the greenhouse effect'.[125] In the 1860s, the British began measuring carbon dioxide, and suggested that the ice ages may have been caused by reduced carbon dioxide in the atmosphere.[126] In the 1890s, the Swedes calculated that increasing concentrations of carbon dioxide in the atmosphere could increase global temperatures by 5°C–6°C.[127] And in the 1940s, British scientists first calculated increasing carbon dioxide from the burning of fossil fuels.[128] Modern science reveals that carbon dioxide concentrations in the atmosphere increased between 1800 and 1900 from 282 ppm to 295 ppm, and by the end of the Second World War (1945) had risen further to 310 ppm.[129] By the 1830s, tropical oceans and the Arctic were beginning to warm.[130] By 1850, Europe, North America, and Asia were warming.[131] By 1945, the global average surface temperature was 0.2°C above pre-industrial levels.[132]

The Cold War Period to the New Millennium

Oil played a substantial and underpinning role in the West's rapid economic recovery from the Second Word War. The US ended its domestic fuel rationing programme causing a boom in the domestic auto industry, with the number of cars on US roads increasing from 26 million to 45 million between 1945 and 1950.[133] In the corresponding period, the US shifted from a net exporter of oil to importing nearly 1 million barrels per day (bpd). To drive Europe's recovery, the US supplied European countries with more than $11 billion in oil aid between 1948 and 1951 under the Marshall Plan. In the coming decades, both the US and Europe would increasingly look to the Middle East to supply their rapidly escalating oil needs.[134] The other victorious economy, the Soviet Union, had fully industrialised by 1950, and had become a world power to rival the US. Between 1955 and 1960, Soviet oil production doubled.[135] However, much of the world had yet to achieve the levels of industrialisation seen in the West, including almost all of Africa and most of Latin America and Asia, including China.[136]

In 1949, the Communist Party leader, Mao Zedong, founded the People's Republic of China. Some scholars describe Mao as 'almost certainly the deadliest individual in history to have wreaked havoc inside a single country'.[137] Researchers contend that the carnage in China under Mao's rule occurred for the same reason it occurred in the Soviet Union under Stalin's rule: to achieve a 'Marxist paradise' in which Capitalists and their offspring were considered especially evil and often put to death for their beliefs.[138] Against this backdrop, three phases of mass atrocities are of particular significance. In phase one, a period of time between 1949 and 1954, Mao undertook a targeted campaign against urban elites (capitalists, westernised intellectuals, Christians, and political opponents) in what Mao's himself acknowledged amounted to 800,000 people being executed – other researchers estimate somewhere 'between one million and four million people'.[139] In addition, at least a further 2.5 million 'class enemies' were dispatched to labour camps with 'conditions no less murderous than the Gulags'.[140] In phase two, between 1958 and 1961, Mao implemented his Great Leap Forward plan. This strategy was supposed to rapidly modernise China's agricultural and industrial systems by emphasising small-scale industries and collective human labour, rather than high technology and centralised factories (the West's model). Mao believed that this alternative approach to industrialisation would propel China past Britain in industrial output.[141] However, not only did this experiment delay China's development by almost 20 years, but it has also been described as causing 'China's worst ever disaster, humanity's worst ever famine, and a death toll numbering in the tens of millions'.[142] Researchers estimate that in 1960 alone, 22 million people died of hunger in China,[143] with the total number of deaths due to famine estimated at 40 million over the three years.[144] Some researchers have described Mao's strategy as 'genocidal' and point to Mao's own words as validation: 'we are prepared to sacrifice 300 million Chinese for the victory of world revolution'.[145] In phase three, between 1966 and 1973, Mao implemented his Great Proletarian Cultural Revolution. This plan has been compared with Stalin's political purges with an estimated 3 million people dying violently.[146]

The 1960s saw fossil fuel-based industrialisation spread around the world and brought with it elevated industrial integration with the West, as accelerated and increasingly interconnected communication, transport, and supply networks combined in a process called 'globalisation'.[147] Western economies, while focusing on new service and technology industries domestically, increasingly sought to outsource 'dirty' industries such as coal-fired manufacturing and extractive operations.[148] It was now possible for Western multinationals to search the globe for a suitable business environment (i.e. lax environmental and labour standards) and set-up shop. Manufacturing boomed in newly industrialised countries in Latin America and Asia as textiles and other industries rushed in, while raw commodities from Africa, the Middle East, the Asia-Pacific, and elsewhere fuelled the production processes. For example, Brazil would develop a strong auto industry with US and European manufacturers seeking cheap labour; and begin an aggressive deforestation epoch (1960–2000) to make way for cattle, soybean, sugarcane (and its derivative – ethanol), and mining enterprises for export to the US especially.[149]

In east Asia, several smaller countries with close ties to the West – particularly the US – began to industrialise on the back of fossil fuels. South Korea would develop an extensive metals and manufacturing industry, overwhelmingly powered coal and oil, and become an increasingly significant global exporter in consumer goods such as electrical appliances, automobiles, and ships.[150] The US provided extensive economic and military aid to South Korea in the aftermath of the Korean War (1950–1953), which saw the South and the US (with a small contingent of allies) fight the North and China (with a small contingent from the USSR). It would cost more than 2.5 million lives, most of them civilians.[151] Taiwan too would soon begin a process of industrialisation and develop a specialisation in exporting cheap consumer goods. In this instance, the US would provide economic and military support to the Chiang Kaishek leadership whose Nationalist Party had only a decade prior been dispatched to the island after more than two decades of large-scale tit for tat massacres with Mao's Communists on mainland China.[152] The British would provide economic and military support to its former colonies, Hong Kong and Singapore, which would help propel them into the industrialised club.[153] Ongoing Western support to post-war Japan would spur a major industrial rebound in that country. The West's concern about the spread of communism served as the backdrop to these industrial developments in Asia.

The 1960s also brought a globally significant campaign urging political separation from the West. Debates about decolonisation dominated in the newly established United Nations (replacing the League of Nations) in the 1950s.[154] Economic inequalities served as a backdrop to a surge of independent movements that led to rapid decolonisation. Such was the global push for decolonisation, UN membership increased from its 51 founding members in

1945 to 100 by 1960 (and to 152 in 1980).[155] Western imperialism, colonisation, and its legacies (such as sharpening economic and ethnic divisions), combined with the acute ideological differences embedded in the Cold War, contributed to the toxic mix that produced some of the most 'conscience-shocking' mass atrocity situations of the century.

For example, in 1954, an armed revolt in Algeria was launched against the French colonial authorities. Hundreds of thousands of lives would be lost in the ensuing war, including many civilians, with notorious state sanctioned torture programmes operating on both sides.[156] In 1962, Algeria declared independence from France. In 1960, Nigeria won its independence from Britain.[157] Soon after, a series of coup attempts would culminate in a civil war between 1967 and 1970 that caused hundreds of thousands of deaths (some estimate 3 million). Massacres and indiscriminate bombing raids were commonplace, but most died from starvation and related causes. The US President Richard Nixon would refer to the Biafran War as a 'genocide'.[158] Indian independence from Britain in 1947 meant drawing up borders for West Pakistan (predominantly Urdu-speaking Muslims), and East Pakistan (predominantly Bengali-speaking Muslims, with a Hindu minority) – now Bangladesh. In 1970, national elections overwhelmingly backed East Pakistan independence from West Pakistan. Unwilling to accept this outcome, in March 1971, the West Pakistani military began massacring Bengalis and Hindus in East Pakistan. Between 1 and 3 million were killed, with sexual violence commonplace.[159]

In the late 1960s and throughout the 1970s, ongoing ideological tensions would continue to provide a backdrop to a string of mass atrocity situations. The US and its allies' belief in the Domino Effect, that countries especially in Asia would gradually fall to Chinese-inspired communism unless forcibly stopped, was a key reason behind the gradual build-up of US military personnel in Vietnam throughout the late 1950s and 1960, peaking in 1969 with close to 550,000 US troops in the country. While atrocities committed during the Vietnam War, like all wars, are legion, the My Lai massacre is the most notable and largest. In this situation, 504 civilians, including children and women, were rounded up and machine gunned to death by US forces, while others were shot in the back as they fled.[160] The indiscriminate use of chemical weapons also became common in this conflict.[161] Vietnam (a former French colony) estimated that around 2 million civilians were killed during the war. US troops officially withdrew from Vietnam in 1973. Communists were also targeted in Indonesia. In this situation, in response to an unsuccessful coup in 1965, the Indonesian military, with civilian support, and backed by the US, slaughtered at least 500,000 people with suspected communist links.[162] Fanning out across the Indonesian archipelago, most victims were beheaded, stabbed, had their throats slit, hacked to death, strangled, hit with clubs, burned alive, or drowned – few were shot.[163]

Communist too would commit mass atrocities. The most egregious occurred between 1975 and 1978 in Cambodia (a former French colony). In this situation, the Khmer Rouge under the leadership of Pol Pot – who like other communist leaders such as Mao and Stalin – exhibited a visceral hatred of 'elites' (wealthy, professionals, educated, and Westerners), ethnic and religious minorities (Vietnamese, Chinese, Muslims), and potential subversives. A campaign of mass executions involving entire families being tortured, shot, and/or bludgeoned to death in 'interrogation centres', combined with internal purges, disease, and famine, would kill close to 2 million Cambodians – one quarter of the country's population – in less than four years, in what has been described as 'one of the worst genocides, relative to population, in recorded history'.[164] The USSR was also directly implicated in multiple mass atrocity situations. For example, the Soviet invasion of Afghanistan in 1979 included the use of chemical weapons, systematic aerial bombardment, and terror tactics, such as gruesome localised atrocities and the destruction of food supplies, that resulted in hundreds of thousands (and up to 1.25 million) Afghan deaths.[165]

By the 1980s, Asia's traditional economic powerhouse Japan was experiencing significant economic growth alongside newcomers, the so-called 'Asian Tigers' (South Korea, Singapore, Taiwan, and Hong Kong). In this decade, Asia's list of industrialised countries would expand to include Malaysia and Thailand that were both engaging significantly in manufacturing. China too would soon get moving. By the late 1970s, China's economy was close to collapse. More than 250 million people in rural China were living in poverty, most on less than a dollar a day, and the economic gap between China and the developed world was growing. In 1978, Deng Xiaoping replaced Mao, who had died in 1976, as Chairman and President of China. The new President immediately launched a strategy of 'Reform and Opening-up' to help spur rapid economic growth. Under the strategy, China's agricultural and heavy industry (and manufacturing) sectors would be gradually exposed to market forces, and China would seek to integrate into the global economy.[166] This more conventional approach to industrialisation, notwithstanding operationalised under strict party control, had the desired effect with factories, roads, and railways proliferating under a double-digit annual growth rate for most of the 1980s (1990s and 2000s).[167] Australia, after decades of cursing its geography, realised that it was in the box seat to fuel Asia's industrial rise with its massive coal and iron ore deposits (and later natural gas), and from the 1970s onwards implemented a series of bipartisan reforms to reorientate its economy away from domestic manufacturing towards mining and minerals export.[168]

In the West, manufacturing outputs grew by 'astonishing rates' between 1960 and 1990: the US growing by 134 per cent, Britain by 195 per cent, Germany by 226 per cent, and France by 323 per cent.[169] Agriculture increasingly used large-scale mechanised equipment and fertilisers. Consumer-driven economic growth was the mantra and was assisted by Western government initiatives and diplomatic leadership to reduce and eliminate tariffs and trade barriers within Europe and globally. Personal ownership of automobiles, televisions, and a variety of appliances became commonplace, and air travel normalised. Computers, robotics, automation, petrochemicals, and plastics (and an emerging renewable energy industry) would become increasingly central to the

West's industrial scene. Resources from the developing world would be pulled into the West's economic orbit to fuel growth, while textiles, auto and steel manufacturing, and coal and other extractive industries would be offshored. Indeed, the high rates of consumption in the West, combined with the economic rise of Asia, meant that industries needed more raw materials than ever before, especially oil.

Since the 1950, the oil-rich countries of the Middle East have been the West's main supplier.[170] Threats to supply have been a constant such as the Suez Canal incident in 1956, the six-day-war in 1967, the Yom Kippur situation in 1973, and the Iranian Revolution in 1979. In 1980, Iran and Iraq went to war, with strategic attacks on oil facilities removing 4 million bpd from the global market. Between 1979 and 1982, US oil imports almost halved causing retail petrol ('gasoline') prices to jump from 63 cents to $1.30 per gallon.[171] The US subsequently increased their military presence in the Persian Gulf to protect oil facilities and shipments. In 1990, Iraq invaded Kuwait following a dispute over an oil field. The US described Iraq's invasion as an economic threat and deployed troops to Saudi Arabia, and subsequently, in 1991, led a multination military mission (with Britain and France, among others) to liberate Kuwait. Atrocities were committed by all parties. For example, Saddam Hussein's notorious Anfal campaign during the Iran–Iraq war involved the extermination of tens of thousands of Kurds in northern Iraq including a chemical attack on the city of Halabja in 1988 that killed approximately 8,000 people in a single day. In 2010, the Iraqi High Criminal Court would describe the campaign as a genocide.[172] Some scholars argue that the UN sanctions regime (spearheaded by the US and UK) imposed on Iraq between 1991 and 2003 exacted such an enormous human toll on Iraqis, especially on children (with up to 500,000 deaths), that it fits within the genocide discussion.[173] Between 1991 and 2000, US retail petrol prices averaged a comparatively low $1.1 a gallon, which, combined with a booming economy and surging demand for SUVs to replace sedans, caused US oil consumption to increase from 16.7 million bpd to 19.7 bpd (a 3-million-barrel per day increase).[174]

In the 1980s and 1990s, several African countries would experience humanitarian tragedies and mass atrocities that would shock the world. In 1956, Sudan gained its independence Britain. The ensuing 'first' civil war between the North and South (1955–1972) cost 500,000 lives; while the second civil war (1983–2005) led to the death of 2.5 million people, and eventual Southern independence.[175] Wrapped up in these conflicts were successive wars and mass atrocity situations in the Darfur region. In neighbouring Ethiopia, two decades of war and atrocities throughout the Tigray region of Ethiopia, combined with a severe and protracted drought, would produce a famine that claimed an estimated 1 million lives between 1983 and 1985. Further South, Uganda's tyrannical President Idi Amin (1971–1979) would commit mass murder and torture throughout his brutal reign, followed by the Ugandan-based militia group, the Lord's Resistance Army (1987–2000s), that was responsible for more than 100,000 deaths, mass rape, mutilations, and the abduction of children across central Africa.[176] In South Africa itself, successive white governments terrorised, murdered, and imprisoned black South Africans en mass under its apartheid regime, eventually collapsing under extreme domestic and international pressure in the 1990s. In West Africa, a civil war in Sierra Leone in which disfigurement, especially chopping off hands, became its hallmark claimed the lives of 200,000 civilians in the 1990s.[177] Child soldiers were commonplace in most of these situations.[178] But a landmark situation not seen since the 1940s would stand above all others.

Scholars have described the Rwandan Genocide as 'probably to most concentrated mass killing ever seen'.[179] On 6 April 1994, a plane carrying majority Hutu President Habyarimana was shot down. Immediately, roadblocks were erected through and around towns, and the extremist Hutu radio station *Mille Collines* declared that 'it's time to cut down the tall trees'. In just 12 weeks in 1994, almost 1 million people – overwhelmingly ethnic minority Tutsis, but also Hutu moderates opposed to the genocide – were massacred with machetes, but also with clubs and small arms, some paying for the privilege to be shot. In one especially egregious incident, on 20 April in the Parish of Karama, between 35,000 and 45,000 people were murdered in less than six hours.[180] The Rwandan genocidaires were not exclusively formal military or informal militia, but also regular citizens – men, women, and children – who hacked to death their neighbours and even friends over this short period of time.[181] The horror ended on 2 June 1994 when the Tutsi-led Rwandan Patriotic Front (RPF) military – led by current Rwandan President, Paul Kagame – took control of key areas. Immediate RPF reprisals and executions amounted to about 50,000 people. In 1996, the RPF invaded east Congo, where most of the genocidaires had fled, and massacred approximately 230,000 people.[182] Rwanda was a Belgium-administrated colony (given to them by Germany after defeat in the First World War), only gaining independence in 1962. Belgium's programme of ethnic favouritism deepened divisions in the country and played a key role in fostering genocidal intent.

Aside from Rwanda, elsewhere in the world, mass atrocities in the 1990s would provide shocking cases of the failure of humanity. In the early 1990, Yugoslavia – a federation of several Slavic groupings – was dissolved. Serbia's president and extreme ethnonationalist, Slobodan Milošević, saw neighbouring Bosnia as a part of a newly formed Greater Serbia, and in 1992 his military invaded, most notably with the sustained encirclement and bombardment of the Bosnian capital, Sarajevo. What ensued was the attempted 'ethnic cleansing' of Muslims from Bosnia. This involved, for the most part, separating the men of 'battle age' – 16 to 60 years old (although younger and older victims were common) – and executing them, while deporting the others. Most dramatically, this policy manifested in Srebrenica in 1995 with 8,000 Bosnian men and boys being rounded up from a UN 'safe haven', loaded into trucks and buses, driven to forests and farms nearby, and machine gunned to death. With the disappearance of men, Serbian soldiers and paramilitary would inflict mass rape on the remaining population. In

1998 and 1999, Serb forces entered Kosovo to 'cleanse' the territory of Albanians through acts of terror and mass murder – about 10,000 Albanians would be killed.[183]

Timor-Leste's (formally East Timor) struggle for independence from Indonesia, which it eventually gained in 1999 after a national referendum, brought to the fore the decades of mass atrocities committed in that country, including the close to 1,400 murders that were committed in the aftermath of the independence vote. It is estimated that between 1975 and 1999, about 170,000 Timorese were killed, including through door-to-door searches and large-scale executions of political opponents (in 1981, for example, an operation resulted in 10,000 executions), and in a particularly depraved situation at Malim Luro in East Timor in 1983, Indonesian troops lined up 60 men, women, and children face down on the ground only to drive over them with a bulldozer.[184] In 1994, Russian forces were sent into Chechnya to overthrow the government that had declared independence from Russia. The massive artillery bombardment, and accompanying ground assaults, that ensued killed tens of thousands of Chechen residents, overwhelmingly civilians.[185] Skirmishes continued in to the 2000s that killed many more Chechen civilians.

By the turn of the millennium 'over half the world was effectively industrialised for the first time'.[186] Oil was the driving force, becoming what historians have called 'the vital fluid of modern industry'.[187] Between 1950 and 2000, global oil consumption had grown by eight times, with oil-based products in 2000 generating around 60 per cent of global energy.[188] Over the same period, global coal consumption had only doubled.[189] Coal's share in global electricity generation hovered at around 40 per cent between 1970 and 2000.[190] However, coal's fortunes would soon change. By the 1990s, China and India – with a combined population of almost 2 billion – began to ramp up their industrialisation plans.[191] Coal would be central to their industrial rise into the new millennium.

Between 1950 and 2000, understanding and awareness of the role that greenhouse gas emissions, especially carbon dioxide, were having on the atmosphere gained pace. In 1957, scientists presented a paper that expressed concern that climate change might be brought about by a build-up of greenhouse gases in the atmosphere through the burning of fossil fuels.[192] In the same year, measurements of carbon dioxide in the atmosphere commenced from an observatory in Hawaii. In 1979, US meteorological satellites began measuring global temperatures.[193] In 1980, the first Villach Conference on climate change brought together leading scientists from a variety of disciplines to share information – the opening lines of the Conference's final statement read:

> World-wide interest and concern have mounted in recent years over an observed increase in carbon dioxide in the atmosphere resulting from human activity. A continuation of this increase is expected to modify the world climate with significant social, economic and political implications.[194]

The second Villach Conference, held in 1985, attended by scientists from 29 developed and developing countries, reported that if carbon dioxide build up in the atmosphere continues at the present rate, climate models show that global surface temperatures could reach between 1.5°C and 4.5°C above pre-industrial times by 2030.[195] These findings informed the Tenth World Meteorological Organization (WMO) conference of 1987 and the landmark report on the environment, *Our Common Future*, published in the same year (see Chapter 2). In 1988, the WMO and UN Environment Programme, with the support of UN Secretary-General Javier Pérez de Cuéllar, established the Intergovernmental Panel on Climate Change (IPCC) to examine and provide timely updates on the science of climate change.[196] In 1990, the IPCC released its First Assessment Report (AR1) on the physical science of climate change. The AR1 was developed by 170 scientists from 25 countries and reviewed by a further 200 scientists.[197] The AR1 and the Second Assessment Report (AR2) published in 1995 continued to confirm that global surface temperatures were rising primarily due to human activity – predominantly through burning fossil fuels. The IPCC's Third Assessment Report (AR3) published in 2001 involved 123 lead authors, 516 contributing authors, 21 review editors, and 420 expert reviewers.[198]

By 2000, it was found that since the Industrial Revolution approximately 600,000 million tonnes of carbon had been emitted into the atmosphere from burning fossil fuels. This had resulted in an increased concentration of carbon dioxide in the atmosphere from 280 ppm to 368 ppm between 1700 and 2000.[199] Climate scientists also reported that global mean surface temperature had risen by about 0.6°C between 1860 and 2000, with the 1990s the warmest decade in last 1000 years.[200] Historians of the Industrial Revolution describe this situation in the following way: 'Climate change seemed to sum-up the most ominous side of industrialisation's latest phase (1950–2000)'.[201]

September 11 to the Covid-19 Pandemic

On September 11, 2001, using tools of globalisation – international communications, finance, and air travel – a group of men launched an attack on two key symbols of Western economic and military power: The World Trade Center and the Pentagon.[202] The attack killed 3,000 people. In response, the US (with British support; and later joined by other Western and local allies) launched airstrikes against Afghanistan – the base of the terrorist group, Al-Qaeda, that had carried out the attack – and subsequent ground assaults. In 2003, the 'war on terror' expanded with the US and coalition forces invading Iraq, toppling the Hussein regime in a matter of months (Hussein later executed in 2006).[203] Skirmishes and suicide attacks in both Afghanistan and Iraq proliferated. In 2021, US forces pulled out of Afghanistan. More than 2,500 US military had been killed,[204] and more than 46,000 civilians killed.[205] In Iraq, more than 4,500 US military personnel had been killed and between 180,000 and 207,000 civilians killed from 2003 to 2011.[206] Tens of thousands more combatants and civilians were injured, many severely debilitated (e.g. losing legs and arms and/or suffering severe brain injuries), in Afghanistan and Iraq over this time.[207]

In December 2010, anti-government protests in Tunisia triggered a wave of uprisings across the Middle East and North Africa (Libya, Egypt, Syria, Yemen, Bahrain, among several other countries) in what has been labelled 'The Arab Spring'. Several governments committed atrocities such as murders and forced disappearances to disperse the protesters (the Libya situation is discussed in several other locations in the present book). Most responses however pale in comparison to the sustained campaign of indiscriminate mass murder actioned against children, women, and men (combatants and civilians) by Syria's President Bashar al-Assad (with Russian air support) between 2011 and 2021. The litany of atrocities committed by the Syrian government include a series of chemical weapons attacks, the worst of which killed approximately 1,400 people, including 400 children – many of them babies and toddlers – in the suburbs of the Ghouta region in August 2013.[208]

In 2014, the Sunni extremist group, Islamic State ('ISIS') – in part a creation of a failed US-led nation building process in Iraq – announced a 'Caliphate' to restore Islamic glory and in the process destroy non-believers.[209] While Shia Muslims were the primary target, ISIS's attempt to exterminate the Yazidis in northern Iraq drew calls of genocide with accounts that 'in village after village, the men and older boys were driven to nearby fields, where they were forced to lie down in the dirt and sprayed with automatic fire'.[210] ISIS affiliates elsewhere in the MENA region, most notoriously Boko Haram operating in West Africa, committed mass atrocities with impunity.[211] Yemen too has become a site where mass atrocities are routinely committed with global indifference with more than 19,000 people, including over 2,300 children, killed or maimed since 2015 by scores of indiscriminate and disproportionate Saudi-led airstrikes (using US and UK munitions).[212]

In Asia, the Myanmar military undertook an extermination campaign against the Rohingya-minority between August and September 2017. Described as a 'textbook example of ethnic cleansing' by the UN, over the two-month period, the Tatmadaw (Myanmar armed forces) shot, stabbed, burned, drowned, and bludgeoned to death Rohingya men, women, and children across Rakhine State – with accounts of infants simply being thrown into a river.[213] It is estimated that more than 25,000 Rohingya were killed, and upward of 700,000 fled across the border to southern Bangladesh. In February 2021, the Myanmar military launched a successful military coup against the leadership of Aung San Suu Kyi (who was in power at the time of the Rohingya genocide). Popular protest opposing the coup prompted the military to use lethal force. As of March 2022, government security forces in Myanmar have killed at least 1,600 people, detained more than 12,500, and burned down more than 10,000 houses.[214]

China's industrialisation (gestating since the 1980s) really took flight in the 2000s. China's comparative advantage was cheap labour and multinationals from around the world – primarily seeking to manufacture inexpensive goods such as toys and appliances – would set up operations to take advantage.[215] In short time, Chinese exports of goods and services boomed, growing from $253 billion to $3.55 trillion between 2000 and 2021.[216] Over the same timeframe, China's GDP per capita soared from $950 to $12,500 per person,[217] along the way lifting more than 200 million people out of poverty (by 2011).[218] To power its factories and produce its 'economic miracle', China primarily burnt coal. China's domestic coal production almost tripped, from 1,355 mega tonnes to 3,764 mega tonnes between 2000 and 2020.[219] In 2019, China accounted for almost half the world's entire coal production, extracting almost 3.7 billion tonnes in that year alone.[220] Coal imports also dramatically increased, growing from 163 million tonnes to 303 million tonnes between 2010 and 2020.[221] Total coal consumption in China quadrupled between 2000 and 2012, reflected in a 50 per cent jump in global coal consumption between 2000 and 2010.[222] By 2021, China consumed more coal than the rest of the world combined.[223] China's consumption of petroleum paralleled this growth, almost quadrupling, from 4.2 million to 15.3 million bpd between 2000 and 2021.[224] The number of cars on Chinese roads doubled, from 50 million to 100 million between 2008 and 2011.[225] Researchers assess that China's rise, particularly since 2000, has produced 'the strongest, most resource-intensive economic growth the world has ever seen'.[226] It also confirmed that fossil fuel-based industrialisation had truly become a global phenomenon.

Understanding among the scientific community about global temperature rise and its causes had become very strong over the period 2000–2022, with the IPCC publishing six new major reports on the science of climate change in this time.[227] For example, the IPCC Fourth Assessment Report (AR4) of 2007, confirmed that 'warming of the climate system is unequivocal', global temperatures have increased by 0.76°C between 1850 and 2005, and 11 of the past 12 years rank among the 12 warmest since the Industrial Revolution.[228] The AR4 also reported that since 1750 global atmospheric concentrations of carbon dioxide have 'increased markedly', primarily as a result of fossil fuel use, to reach 379 ppm in 2005. The AR4 concluded that 'There is very high confidence that the net effect of human activities since 1750 has been one of warming'.[229] The IPCC Sixth Assessment Report (AR6), released in 2021, continued to report on the upward temperature and emissions trajectories. For example, it found that the 'global surface temperature was 1.09°C higher' in 2020 compared to 1850, and 'each of the last four decades have been successively warmer than any decade that preceded it since 1850'.[230] Concentrations of carbon dioxide, it added, have continued to increase in the atmosphere, reaching 410 ppm in 2019.[231] The increase of carbon dioxide in the atmosphere since the Industrial Revolution is 'unequivocally caused by human activities', it added.[232] In January 2023, concentrations of carbon dioxide reached 419 ppm.[233]

Any number of indicators could be included in this discussion as further evidence of the negative impacts of human activity on the climate system.

For example, the concentration of methane in the atmosphere, a gas primarily produced through livestock that is 20 times more damaging than carbon dioxide, has more than doubled since pre-industrial times, reaching over 1,800 parts per billion in recent years.[234] Artic sea ice has shrunk by more than 2 million square km since satellite observations began in 1979.[235] The last 10 years were the ocean's warmest decade since at least the 1800s.[236] As a consequence of ice sheet and glacial melt and thermal expansion, global sea level has increased by 200 millimetres since 1900, with half of that rise (100mm) recorded between 1995 and 2020.[237] Half of the world's tropical and temperate forests are now gone.[238] The number of reported climate-related disasters (drought, extreme temperatures, flood, landslides, storms, and wildfires) have increased more than four-fold over the past 50 years, from 711 in 1970–1979 to 3165 in 2010–2019.[239] And economic losses due to climate impacts have increased by a factor of seven over the same period, from $175 billion in 1970–1979 to $1.3 trillion in 2010–2019.[240] In short, all the key indicators on climate change (warming, causing, and impacts) are heading in the wrong direction.

Despite the carnage visited upon large swathes of the Middle East and Africa, and parts of Asia (discussed above), researchers have found that the period from 2000 (and even since the 1990s) to 2010 was one of relative global peace, popularised with titles such as 'The Better Angels of Our Nature', and 'Winning the War on War'.[241] Indicators of armed conflict, civilian casualties, atrocity impunity, and forced displacement tend to demonstrate this claim.[242]

However, since the Arab Spring uprisings in 2011, these key atrocity indicators have all moved in the opposite direction, in what has been titled, 'The Return of History'.[243] For example, the number of major civil wars (those with over 1,000 battle deaths per year and involving at least one state actor) tripled from 4 in 2011 to 12 in 2020 (in Afghanistan, Iraq, Myanmar, Nigeria × 2, Pakistan, Somalia, Sudan, Syria × 2, Ukraine, and Yemen).[244] Violence against civilians (forced disappearance, attack, and sexual violence) increased from about 1,500 events annually between 2000 and 2009, before climbing to 4,700 in 2015 and to nearly 29,000 in 2021.[245] The associated fatalities show a similar trend from a low point of 2,100 deaths in 2006, before rising sharply after 2011 to almost 34,000 deaths in 2021.[246] The abuse of children in the context of armed conflict also appears to have increased.[247] For example, in 2015, the UN identified 54 'parties' (armed forces and groups) globally that had engaged in activities targeting children (killing, maiming, recruitment, sexual violence, abduction, attacks on schools and hospitals), which represented the highest number since reporting began in 2003.[248] And the number of people forcibly displaced by war, violence, persecution, and human rights abuses according to the United Nations High Commissioner for Refugees (UNHCR) hovered at around 40 million annually between 1990 to 2011, before sharply increasing to 89 million by 2021.[249]

Atrocity situations continue to the present time across the world including in Mali, Niger, Sudan, South Sudan, Democratic Republic of Congo, Central African Republic,[250] and, most notably, across the Tigray region of Ethiopia where 'acts of ethnic cleansing' were committed in 2021, according to the US government.[251] Russia's invasion of Ukraine in February 2022 has resulted in accusations that Russian soldiers have committed war crimes after several mass graves were uncovered.[252] And as China looks to further expand militarily in the Asia-Pacific, claims of genocide have been levelled against the Chinese government for their ongoing systematic persecution of Uyghurs in Xinjiang province.[253]

Conclusion

Ultimately, this chapter showed that the parallel problems of climatic change (and emissions – post 1760) and atrocities have always occurred, but of late are getting progressively worse. It also suggested that there appears to be a complex relationship involving climate change, between the rise and fall of economic power and atrocity situations. The first section, for example, showed that natural climate variability shaped the economic and political circumstances of the Mayans which contributed to civilisational disintegration and an increase in atrocity events. The second section showed that the West's fossil fuel-based industrialisation and subsequent internal economic competition, and by implication military competition, encouraged imperialist policies that resulted in mass atrocities the world over. It also showed that fossil fuels (their access and use) were central to the imperialist visions that drove the two major world wars in this period. The third section showed that the ideological struggle between communist and capitalist systems spilled over into visceral hatred, warfighting, and numerous atrocity situations. The fourth section showed that coal and oil continue to be deeply intertwined with the global economic and military balance of power – changes in which appear to almost always have grave implications for men, women and children just trying to live their lives.

The next chapter tracks the major international policy efforts that have sought to curb human-caused global warming and prevent mass atrocities.

Notes

1 Beatrice Mosello, Lukas Rüttinger, and Liesa Sauerhammer, 'The Climate Change-Conflict Connection – The Current State of Knowledge', *Adelphi*, 2019. See also, Thomas F. Homer-Dixon, 'Environmental Scarcities and Violent Conflict: Evidence from Cases', *International Security*, Vol 19, No 1, Summer, 1994, p 5–40. Sherri Goodman and Pauline Baudu, 'Climate Change as a "Threat Multiplier": History, Uses and Future of the Concept', *Centre for Climate and Security*, 03 January 2023. See also the innovative 'pathways approach' between climate, conflict, peace, and security in Florian Krampe, 'Pathways of Climate Insecurity: Guidance for Policymakers', *SIPRI*, November 2020.

2 For other examples of a foreground and background distinction, see for example, Stella Schaller, Katarina Schulz, and Beatrice Mosello (and editorial team: Benjamin Pohl, Janani Vivekananda, Adrian Foong, Emily Wright, Daria Ivleva, Pia van

Ackern, Adrien Detges, Lukas Rüttinger, Jonathan R. Smith), 'Climate Change and Security: The Handbook', Climate Diplomacy and Adelphi, May 2020, p 8.

3 David M. Crowe, *War Crimes, Genocide and Justice: A Global History* (New York: Palgrave Macmillan, 2014), p 8.

4 Michael Bryant, *A World History of War Crimes: From Antiquity to the Present* (London and New York: Bloomsbury, 2016), p 11.

5 Ibid.

6 Ibid., p 12.

7 E. Jansen, J. Overpeck, K.R. Briffa, J.-C. Duplessy, F. Joos, V. Masson-Delmotte, D. Olago, B. Otto-Bliesner, W.R. Peltier, S. Rahmstorf, R. Ramesh, D. Raynaud, D. Rind, O. Solomina, R. Villalba, and D. Zhang, 'Palaeoclimate' in *Climate Change 2007: The Physical Science Basis*. Contribution of Working Group I to the Fourth Assessment Report of the Intergovernmental Panel on Climate Change (Cambridge: Cambridge University Press, 2007).

8 Ben Kiernan, *Blood and Soil: A World History of Genocide and Extermination from Sparta to Darfur* (New Haven and London: Yale University Press, 2007), p 45.

9 Adam Jones, *Genocide: A Comprehensive Introduction*, 3rd Edition (London: Routledge, 2017), p 7.

10 Ibid., p 4.

11 Crowe, *War Crimes, Genocide and Justice*, p 13–14.

12 Jones, *Genocide: A Comprehensive Introduction*, p 8.

13 IPCC, 2014: *Climate Change 2014: Impacts, Adaptation, and Vulnerability. Part A: Global and Sectoral Aspects.* Contribution of Working Group II to the Fifth Assessment Report of the Intergovernmental Panel on Climate Change, (Cambride and New York: Cambridge University Press. 2014), p 772. For a broad analysis of the preindustrial period see Karl W. Butzer, 'Collapse, Environment, and Society', *PNAS*, Vol 109, No 10, March 2012, p 3632–3639.

14 Harvey Weiss (Eds.), *Megadrought and Collapse: From Early Agriculture to Angkor* (New York: Oxford University Press, 2017), p 1. Weiss examines nine megadroughts – which he defines as 'severe periods of drought lasting decades or centuries' – and societal collapse events from the Holocene to 1500 AD.

15 Jared Diamond, *Collapse: How Societies Choose to Fail or Succeed* (New York: Penguin Group, 2005); Butzer, 'Collapse, Environment, and Society'. p 3632–3639.

16 It should be noted that other major cases of civilizational collapse involving climate change did not result in an uptick in atrocity events – for example, the economic and political collapse of Angkor (present-day Cambodia). See, for example, Brendan M. Buckley, et al., 'Climate as a Contributing Factor in the Demise of Angkor, Cambodia', *PNAS*, Vol 107, No 15, 2010, p 6748–6752; and Roland Fletcher, Brendan M. Buckley, Christophe Pottier, and Shi- Y u Simon Wang, 'Fourteenth to Sixteenth Centuries AD: The Case of Angkor and Monsoon Extremes in Mainland Southeast Asia' in Weiss, *Megadrought and Collapse.*

17 Douglas J. Kennett and David A. Hodell, 'AD 750–1100 Climate Change and Critical Transitions in Classic Maya Socio-political Networks', in Weiss, *Megadrought and Collapse*, p 219. See also David Webster, 'The Not So Peaceful Civilization: A Review of Maya War', *Journal of World Prehistory*, Vol 14, No 1, 2000, p 65–199.

18 W. Christopher Carleton, David Campbell, and Mark Collard, 'Increasing Temperature Exacerbated Classic Maya Conflict Over the Long Term', *Quaternary Science Reviews*, Vol 163, 2017, p 209–218. See also, Douglas J. Kennett et al.,

'Development and Disintegration of Maya Political Systems in Response to Climate Change', *Science*, Vol 338, No 6108, 2012, p 791.
19 Webster, 'The Not So Peaceful Civilization', p 65–119.
20 Carleton, Campbell, and Collard, 'Increasing Temperature Exacerbated Classic Maya Conflict', p 209–218.
21 Webster, 'The Not So Peaceful Civilization', p 65–119.
22 Takeshi Inomata, 'The Last Day of a Fortified Classic Maya Center: Archaeological Investigations at Aguateca, Guatemala', *Ancient Mesoamerica*, Vol 8, No 2, Fall 1997, p 348.
23 'Archaeologists in Peru Unearth 227 Bodies in Biggest-ever Discovery of Child Sacrifice', *ABC News*, 29 August 2019.
24 Kennett and Hodell, 'AD 750–1100 Climate Change and Maya Socio-political Networks', in Weiss, *Megadrought and Collapse*, p 217.
25 Brian Fagan, *The Great Warming: Climate Change and the Rise and Fall of Civilizations* (New York: Bloomsbury Press, 2008), p 11 and 18.
26 Jones, *Genocide: A Comprehensive Introduction*, p 9.
27 Ibid.
28 Ibid., p 8.
29 Fagan, *The Great Warming*, p 51.
30 Ibid., p 11 and 18.
31 Ibid.
32 Ibid., p 12.
33 Ibid., p 24–25.
34 Brian Fagan, *The Little Ice Age: How Climate Made History 1300 –1850* (New York: Basic Books, 2000).
35 Geoffrey Parker, 'Crisis and Catastrophe: The Global Crisis of the Seventeenth Century Reconsidered', *The American Historical Review*, Vol 113, No 4, 2008, p 1068.
36 W. Neil Adger, et al., 'Human Security', in IPCC, 2014: *Climate Change 2014: Impacts, Adaptation, and Vulnerability. Part A: Global and Sectoral Aspects*. Contribution of Working Group II to the Fifth Assessment Report of the Intergovernmental Panel on Climate Change (Cambridge and New York: Cambridge University Press, 2014), p 771 and 772.
37 See for instance, Timothy Brook, *The Troubled Empire: China in the Yuan and Ming Dynasties* (Harvard University Press, 2013); and Richard S. J. Tol and Sebastian Wagner, 'Climate Change and Violent Conflict in Europe Over the Last Millennium', *Climatic Change*, Vol 99, 2010, p 65–79.
38 See for example, David D. Zhang, C. Y. Jim, George C-S Lin, Yuan-Qing He, James J. Wang and Harry F. Lee, 'Climatic Change, Wars and Dynastic Cycles in China: Over the Last Millennium', *Climatic Change*, Vol 76, 2006, p 459–477.
39 Ibid.
40 Jones, *Genocide: A Comprehensive Introduction*, p 8.
41 Tol and Wagner, 'Climate Change and Violent Conflict in Europe', p 67.
42 Parker, 'Crisis and Catastrophe', p 1056.
43 Thomas Hobbes, *Leviathan* (1651), p xiii.
44 David D. Zhanga, Harry F. Lee, Cong Wang, Baosheng Li, Qing Peia, Jane Zhang, and Yulun An, 'The Causality Analysis of Climate Change and Large-Scale Human Crisis', *PNAS*, Vol 108, 2011, p 17301.
45 Parker, 'Crisis and Catastrophe', p 1073.

46 Ibid., p 1073. See also, Adger, et al., 'Human Security', in IPCC, 2014, p 772.
47 Le Roy Ladurie in Parker, 'Crisis and Catastrophe', p 1077.
48 Fagan, *The Little Ice Age*, p xvi.
49 Jones, *Genocide: A Comprehensive Introduction*, p 151.
50 Ibid., p 8.
51 Ibid., p 156.
52 Ibid., p 157.
53 Peter Stearns, *The Industrial Revolution in World History*, 4th Edition (Boulder: Westview Press, 2013), p 31. See also, John U. Nef, 'An Early Energy Crisis and Its Consequences', *Scientific America*, November 1977.
54 From the Editor, 'When Britain Ran Out of Wood', *Challenger Magazine*, Vol 3, 1979.
55 John Houghton, *Global Warming: The Complete Briefing*, 3rd Edition (New York: Cambridge University Press, 2004), p 65.
56 Ibid., p 29.
57 Ibid., p 65.
58 Stearns, *The Industrial Revolution in World History*, p 6.
59 Ibid., p 7.
60 Ibid., p 30.
61 Ibid., p 31.
62 Ibid., p 37.
63 Ibid., p 37.
64 'It's All About Steam', *The Transcontinental Railroad* (accessed 2022).
65 'Transatlantic Slave Trade', *Britannica* (accessed 2022).
66 Jones, *Genocide: A Comprehensive Introduction*, p 52.
67 Stearns, *The Industrial Revolution in World History*, p 53.
68 Herman van der Wee, 'The Industrial Revolution in Belgium', in Mikulas Teich and Roy Porter (Eds.), *The Industrial Revolution in National Context: Europe and the USA* (Cambridge: Cambridge University Press, 1996), p 72.
69 Joel Mokyr, 'The Industrial Revolution in the Low Countries in the First Half of the Nineteenth Century: A Comparative Case Study', *The Journal of Economic History*, Vol 34, No 2, 1974, p 366.
70 Jones, *Genocide: A Comprehensive Introduction*, p 10.
71 Stearns, *The Industrial Revolution in World History*, p 57–58.
72 Ibid., p 61–66.
73 Ibid., p 13.
74 Jones, *Genocide: A Comprehensive Introduction*, p 157.
75 Ibid., p 160.
76 'Oil Dependence and U.S. Foreign Policy: 1950–2022', *Council on Foreign Relations* (accessed 2022).
77 Stearns, *The Industrial Revolution in World History*, p 66 and 57.
78 Ibid., p 68.
79 Ibid.
80 Ibid., p 13, 53, and 104.
81 M. E. Chamberlain, *The Scramble for Africa*, 3rd Edition (London: Routledge, 2013). See also, Ewout Frankema, Jeffrey Williamson, and Pieter Woltjer, 'An Economic Rationale for the West African Scramble? The Commercial Transition and the Commodity Price Boom of 1835–1885', *The Journal of Economic History*, Vol 78, No 1, p 231–267.

82 Jones, *Genocide: A Comprehensive Introduction*, p 8.
83 Ibid., p 10.
84 Ibid., p 166.
85 J. P. Daughton, 'Behind the Imperial Curtain: International Humanitarian Efforts and the Critique of French Colonialism in the Interwar Years', *French Historical Studies*, Vol 34, No 3, 2011, p 515.
86 Kiernan, *Blood and Soil*, p 364.
87 Stearns, *The Industrial Revolution in World History*, p 177.
88 Ibid., p 127.
89 Ibid.
90 Ibid., p 123.
91 Ibid., p 132–133.
92 Ibid., p 128.
93 Jones, *Genocide: A Comprehensive Introduction*, p 259–260.
94 Stearns, *The Industrial Revolution in World History*, p 161.
95 'Oil Dependence and U.S. Foreign Policy (accessed 2022).
96 Ibid.
97 Ibid.
98 Stearns, *The Industrial Revolution in World History*, p 162.
99 Ibid., p 173.
100 Jones, *Genocide: A Comprehensive Introduction*, p 122. See also for an excellent account of the situation, Martin Middlebrook, *The First Day on the Somme* (London: Penguin Books, 2016).
101 Jones, *Genocide: A Comprehensive Introduction*, p 200–202.
102 'Oil Dependence and U.S. Foreign Policy', (accessed 2022).
103 Bogdan Costea and Peter Watt, 'How a Soviet Miner from the 1930s Helped Create Today's Intense Corporate Workplace Culture', *The Conversation*, 29 June 2021.
104 Stearns, *The Industrial Revolution in World History*, p 136–138.
105 Jones, *Genocide: A Comprehensive Introduction*, p 261.
106 Stearns, *The Industrial Revolution in World History*, p 137.
107 Jones, *Genocide: A Comprehensive Introduction*, p 263.
108 Ibid., p 264.
109 Ibid., p 270.
110 Ibid., p 124.
111 Ibid.
112 Ibid.
113 Ibid., p 126.
114 Ibid., p 274.
115 Ibid., p 55 and 383.
116 Ibid., p 325.
117 Ibid., p 332.
118 Ibid., p 318.
119 Ibid., p 100.
120 Ibid.
121 Ibid.
122 Ibid., p 53 and 55–57.
123 'Oil Dependence and U.S. Foreign Policy', (accessed 2022).
124 Houghton, *Global Warming*, p 17.
125 Ibid.

126 Ibid.
127 Ibid.
128 Ibid.
129 'Trends in Atmospheric Concentrations of CO2 (ppm), CH4 (ppb) and N2O (ppb), between 1800 and 2017', *European Environment Agency* (accessed 2022). See also Johannes Friedrich and Thomas Damassa, 'The History of Carbon Dioxide Emissions', *World Resources Institute*, 21 May 2014.
130 Helen McGregor, Joelle Gergis, Nerilie Abram, and Steven Phipps, The Industrial Revolution Kick-started Global Warming Much Earlier than we Realised, *The Conversation*, 2016.
131 Ibid.
132 'World of Change, Global Temperatures', Earth Observatory, *NASA* (accessed 2022).
133 'Oil Dependence and U.S. Foreign Policy', (accessed 2022).
134 Stearns, *The Industrial Revolution in World History*, p 204.
135 Sergei Ermolaev, 'The Formation and Evolution of the Soviet Union's Oil and Gas Dependence', Working Paper, *Carnegie Endowment for International Peace*, 29 March 2017, p 136–138.
136 Stearns, *The Industrial Revolution in World History*, p 112 and 177.
137 Jones, *Genocide: A Comprehensive Introduction*, p 282.
138 Ibid.
139 Ibid., p 284.
140 Ibid., p 285.
141 Stearns, *The Industrial Revolution in World History*, p 223.
142 Jones, *Genocide: A Comprehensive Introduction*, p 284.
143 Ibid., 290.
144 Yang Jisheng, *Tombstone: The Great Chinese Famine 1958–1962* (New York: Farrar Strauss & Giroux, 2013).
145 Jones, *Genocide: A Comprehensive Introduction*, p 290.
146 Ibid., p 293. Mao also invaded and put down several rebellions in Tibet and elsewhere that involved massacres of tens of thousands – see Jones, *Genocide: A Comprehensive Introduction*, p 287.
147 Stearns, *The Industrial Revolution in World History*, p 207 and 202.
148 Ibid., p 207.
149 Ben Parr, Carlos Eduardo F. Young, and Don Henry, 'Opportunities for a Low Pollution Economic Recovery in Brazil', Briefing Paper, University of Melbourne, May 2017.
150 John Byrne, Leigh Glover, Hoesung Lee, Young-DooWang, and Jung-Min Yu, 'Electricity Reform at a Crossroads: Problems in South Korea's Power Liberalization Strategy', *Pacific Affairs,* Vol 77, No 3, 2004, p 496.
151 Stearns, *The Industrial Revolution in World History*, p 215. See also, Dong Choon Kim, 'Forgotten War, Forgotten Massacres—the Korean War (1950–1953) as Licensed Mass Killings', *Journal of Genocide Research*, Vol 6, No 4, 2004, p 523–544.
152 Stearns, *The Industrial Revolution in World History*, p 214.
153 Ibid., p 203.
154 David Reynolds 'Empire, Region, World: The International Context of Australian Foreign Policy Since 1939', *Australian Journal of Politics and History*, Vol 51, No. 3, 2005, p 349.

155 Ibid.
156 Bryant, *A World History of War Crimes*, p 200–201.
157 Michael Gould, *The Biafran War: The Struggle for Modern Nigeria* (London and New York: I.B. Tauris, 2013).
158 Jones, *Genocide: A Comprehensive Introduction*, p 58.
159 Ibid., p 457–466.
160 Ibid., p 103.
161 Ibid., p 104.
162 Ibid., p 419.
163 Ibid., p 421.
164 Jones, *Genocide: A Comprehensive Introduction*, p 402 and 258 and 406.
165 Ibid., p 107.
166 Ben L. Parr and Don Henry, 'A New Starting Point: China's Eco-civilisation and Climate Action Post-Paris', Briefing Paper, University of Melbourne, June 2016.
167 Jones, *Genocide: A Comprehensive Introduction*, p 284.
168 Ben L. Parr, *Australian Climate Policy and Diplomacy: The Transition Years, John Howard to Kevin Rudd*, Doctoral Thesis, University of Melbourne, 2015, p 124.
169 Stearns, *The Industrial Revolution in World History*, p 240.
170 Ibid. p 232.
171 'Historical Gasoline Prices, 1929–2011', *US Department of Energy* (accessed 2022).
172 Bahar Baser and Mari Toivanen, 'The Politics of Genocide Recognition: Kurdish Nation-building and Commemoration in the post-Saddam Era', *Journal of Genocide Research*, Vol 19, No 3, 2017, p 409.
173 Jones, *Genocide: A Comprehensive Introduction*, p 59–60.
174 'Oil and Petroleum Products Explained: Oil Imports and Exports', *US Energy Information Administration* (accessed 2022).
175 Andrew S. Natsios, *Sudan, South Sudan and Darfur: What Everyone Needs to Know* (New York: Oxford University Press, 2012), p 1–3.
176 'Central Africa (The Lord's Resistance Army)', *Global Centre for the R2P* (accessed 2022).
177 'Shocking War Crimes in Sierra Leone: New Testimonies on Mutilation, Rape of Civilians', *Human Rights Watch*, 24 June 1999. See also, 'Sierra Leone: 50-Year Sentence for Charles Taylor', *Human Rights Watch*, 30 May 2012.
178 See, for example, Romeo Dallaire, *They Fight Like Soldiers, They Die Like Children* (Arrow Books, 2011).
179 Jones, *Genocide: A Comprehensive Introduction*, p 354. See also, Romeo Dallaire, *Shake Hands with the Devil: The Failure of Humanity in Rwanda* (Toronto: Random House, 2003); and Philip Gourevitch, *We Wish to Inform You that Tomorrow We Will Be Killed with Our Families: Stories from Rwanda* (New York: Picador, 1999).
180 Jones, *Genocide: A Comprehensive Introduction*, p 482.
181 Ibid., p 473.
182 Ibid., p 490.
183 Ibid., p 445.
184 Ibid., p 425.
185 Ibid., p 312.
186 Stearns, *The Industrial Revolution in World History*, p 202.
187 Ibid., p 233.
188 Hannah Ritchie, Pablo Rosado, and Max Roser, 'Fossil Fuels', *Our World in Data* (accessed 2022).

189 Ibid.
190 'World Energy Balances: Overview', *International Energy Agency* (accessed 2022).
191 Stearns, *The Industrial Revolution in World History*, p 202.
192 Houghton, *Global Warming*, p 17.
193 Ibid., p 59. See also, 'The Origins of the IPCC: How the World Woke up to Climate Change, *International Science Council* (accessed 2022).
194 'Review of the Global Environment 10 Years After Stockholm', Hearings before the Subcommittee on Human Rights and International Organizations of the Committee on Foreign Affairs, House of Representatives, Ninety-seventh Congress, Second Session, 1982, p 119.
195 ICSU, UNEP, WMO: Report of the International Conference on the Assessment of the Role of Carbon Dioxide and of Other Greenhouse Gases in Climate Variations and Associated Impacts, Villach, Austria, 9–15 October 1985, p 1.
196 In addition, NASA scientist James Hanses would testify to US congress that fossil fuels are causing temperatures to rise.
197 Houghton, *Global Warming*, p 220.
198 Ibid., p 219–220.
199 Ibid., p 31 and 62.
200 Ibid., p 62 and 56.
201 Stearns, *The Industrial Revolution in World History*, p 208.
202 John Baylis, Patricia Owens, and Steve Smith (Eds.), *The Globalization of World Politics: An Introduction to International Relations*, 4th Edition (New York: Oxford University Press, 2008), p 2.
203 'The Iraq War: 2003–2011', *Council on Foreign Relations* (accessed 2022).
204 Neta C. Crawford, 'Calculating the Costs of the Afghanistan War in Lives, Dollars and Years', *The Conversation*, 2 September 2021.
205 See 'Costs of War (Human Costs of Post-9/11 Wars)', Watson Institute, *Brown University* (accessed 2022).
206 Ibid.
207 'Reports on the Protection of Civilians in Armed Conflict', United Nations Assistance Mission in Afghanistan (accessed 2022).
208 Alex J. Bellamy, 'The Responsibility to Protect at 15: A Promise Unfulfilled', *Global Centre for the R2P*, 21 September 2020.
209 Jones, *Genocide: A Comprehensive Introduction*, p 250.
210 Ibid., p 253.
211 'Nigeria', *Global Centre for the R2P* (accessed 2022). Anietie Ewang, 'Gruesome Boko Haram Killings in Northeast Nigeria: Authorities Should Prioritize Civilian Protection', *Human Rights Watch*, 1 December 2020.
212 See chapter on Yemen in 'World Report, 2019: Events of 2018', *Human Rights Watch*, 2019; 'Yemen', Global Centre for the R2P (accessed 2022).
213 'UN Human Rights Chief Points to 'Textbook Example of Ethnic Cleansing' in Myanmar', *UN News*, 11 September 2017. See also, 'Report of the Detailed Findings of the Independent International Fact-Finding Mission on Myanmar', *UN Human Rights Council*, 12 September 2018. In 2022, the US government classified the situation as a genocide.
214 'Myanmar: 'Appalling' Violations Demand 'Unified and Resolute International Response', *UN News*, 15 March 2022.
215 Stearns, *The Industrial Revolution in World History*, p 225.
216 'Exports of Goods and Services – China', *The World Bank* (accessed 2022).

217 'GDP Per Capita – China', *The World Bank* (accessed 2022).
218 Tony Saich, *Governance and Politics of China*, 4th Edition (Palgrave Macmillian, 2015), p 305.
219 'China', *International Energy Agency* (accessed 2022); 'Coal Production by Country', *Our World in Data* (accessed 2022).
220 Andrew Fawthrop, 'Profiling the Six Largest Coal-Producing Countries Around the World', *NS Energy*, 4 January 2021.
221 'Import Volume of Coal to China Between 2010 and 2020', *Statista* (accessed 2022).
222 'World Coal Consumption, 1978–2020', *International Energy Agency* (accessed 2022).
223 'Coal Consumption by Country', *WorldoMeter* (accessed 2022). 'China Produces and Consumes Almost as Much Coal as the Rest of the World Combined', *US Energy Information Energy*, 14 May 2014.
224 'Country Analysis Executive Summary: China', *US Energy Information Energy*, 8 August 2022.
225 'The Cost of Air Pollution: Health Impacts of Road Transport', *OECD*, 2014, p 4.
226 Ross Garnaut, 'The New Model of Growth and the Global Resources Economy', in Ligang Song, Ross Garnaut, Cai Fang, and Lauren Johnston (Eds.), *China's Domestic Transformation in a Global Context* (Canberra: ANU Press, 2015), p 19.
227 Special Report on Emissions Scenarios, 2000; Third Assessment Report, 2001; Fourth Assessment Report, 2007; Fifth Assessment Report, 2014; Special Report on Global Warming of 1.5C, 2018; and Sixth Assessment Report, 2022
228 IPCC, 2007: *Climate Change 2007: Synthesis Report.* Contribution of Working Groups I, II and III to the Fourth Assessment Report of the Intergovernmental Panel on Climate Change (Cambridge: Cambridge University Press, 2007), p 2.
229 Ibid., p 5.
230 IPCC, 2021: *Summary for Policymakers.* In: *Climate Change 2021: The Physical Science Basis. Contribution of Working Group I to the Sixth Assessment Report of the Intergovernmental Panel on Climate Change* (Cambridge: Cambridge University Press, 2021), p 5.
231 Ibid., p 4.
232 Ibid.
233 'Climate Change: Vital Signs of the Planet', *NASA* (accessed 2023).
234 'Climate Change Indicators: Atmospheric Concentrations of Greenhouse Gases', *US Environmental Protection Agency* (accessed 2023).
235 'Global Climate Change', *NASA* (accessed 2023).
236 Ibid.
237 Ibid.
238 For an extended list of environmental indicators, see Peter Newell, *Globalization and The Environment: Capitalism, Ecology and Power* (Cambridge: Polity Press, 2012), p 2.
239 *WMO Atlas of Mortality and Economic Losses from Weather, Climate and Water Extremes (1970–2019)* (Geneva: World Meteorological Organization, 2021), p 16–19. See also, James MacCarthy, Sasha Tyukavina, Mikaela Weisse, and Nancy Harris, 'New Data Confirms: Forest Fires Are Getting Worse', *World Resources Institute*, 17 August 2022.
240 *WMO Atlas of Loses from Climate (1970–2019)* p 16–19. See also, 'The Rising Costs of Extreme Weather Events', *The White House*, 1 September 2022.

241 Steven Pinker, *The Better Angels of Our Nature: Why Violence Has Declined* (New York: Penguin Books, 2011). Joshua Goldstein, *Winning the War on War: The Decline of Armed Conflict Worldwide* (New York: Dutton, 2011).
242 Erik Melander, Therése Pettersson, and Lotta Themnér, 'Orgnaised Violence, 1989–2015', *Journal of Peace Research*, Vol 53, No 5, 2016, 727–742 (Uppsala Conflict Data Program, Uppsala University). See also, Thomas S. Szayna, Stephen Watts, Angela O'Mahony, Bryan Frederick, and Jennifer Kavanagh, 'What Are the Trends in Armed Conflicts, and What Do They Mean for U.S. Defense Policy?', *RAND*, 2017.
243 Jennifer Welsh, *The Return of History: Conflict, Migration and Geopolitics in the Twenty-First Century* (Toronto: House of Anansi Press, 2017). See also, Alex J. Bellamy and Ivan Simonovic, 'Introduction: Towards Evidence Based Atrocity Prevention', *Journal of International Peacekeeping*, 24, 2020, p 285–304; See also, Bellamy, 'The Responsibility to Protect at 15'.
244 Bellamy and Simonovic, 'Towards Evidence Based Atrocity Prevention', p 294. See also, Sebastian von Einsiedel, 'Civil War Trends and the Changing Nature of Armed Conflict', *United Nations University Centre for Policy Research*, March 2017; Max Roser, Joe Hasell, Bastian Herre, and Bobbie Macdonald, 'War and Peace', *Our World in Data* (accessed 2022); Havard Strand, Siri Aas Rustad, Henrik Urdal, and Havard Mokleiv Nygard, 'Trends in Armed Conflict, 1946–2018', Conflict Trends, Vol 3, *PRIO*, 2019.
245 'ACLED' (accessed 2022).
246 Ibid. See also, 'Ninety Per Cent of War-Time Casualties Are Civilians, Speakers Stress, Pressing Security Council to Fulfil Responsibility, Protect Innocent People in Conflicts', Meeting Coverage, Security Council, *United Nations*, 25 May 2022.
247 von Einsiedel, 'Civil War Trends'.
248 'Children and Armed Conflict', Report of the Secretary-General (A/70/836–S/2016/360), 20 April 2016, p 37–39. See also, von Einsiedel, 'Civil War Trends', p 7–8.
249 'Global Trends Forced Displacement in 2021', *UNHCR*, 2022, p 2. See also, 'UNHCR: Global Displacement Hits Another Record, Capping Decade-long Rising Trend', *UNHCR*, 16 June 2022; 'Atrocity Alert No. 258: Special Issue on Global Displacement and Mass Atrocities', *Global Centre for R2P*, 23 June 2021; 'Global Displacement Forecast: 2022', *Danish Refugee Council*, February 2022; 'Forced Displacement: Refugees, Internally Displaced and Host Communities', *The World Bank* (accessed 2022); Bellamy, 'The Responsibility to Protect at 15'.
250 See, for example, 'Populations at Risk', *Global Centre for the R2P* (accessed 2022)
251 'Blinken: Acts of 'Ethnic Cleansing' Committed in Western Tigray', *CNN*, 10 March 2021; 'Ethiopia's Invisible Ethnic Cleansing: The World Can't Afford to Ignore Tigray', *Human Rights Watch*, 16 June 2022.
252 'Ukraine: Apparent War Crimes in Russia-Controlled Areas Summary Executions, Other Grave Abuses by Russian Forces', *Human Rights Watch*, 3 April 2022. 'Some Bodies Found at Mass Burial Site in Izium Show 'signs of Torture,' Ukraine Says', *CNN*, September 17, 2022.
253 'China', *Global Centre for the R2P* (accessed 2022).

2 Efforts to Curb Global Warming and Prevent Atrocities

This chapter builds on the previous chapter by historicising the major international policy efforts to curb global warming and prevent atrocities. As is the core feature of the present book, the chapter traces each history in parallel, but where noticeable, identifies linkages between both agendas. Ultimately, the chapter contends that the background (structural) aim of both policy development processes was, and is, to protect vulnerable human populations from large-scale harm. However, the overall timeframe examined reveals several instances (moments, stands, and periods) of foreground divergence as well as convergence.

The first section covering the period from the end of the Cold War to the World Trade Center attacks (1989–2001), shows that a closeness existed between the atrocity prevention and climate action agendas. For the most part, this closeness manifested as each agenda identifying the other as providing further impetus for enhanced policy action in their separate policy domains. However, two major points of convergence can be identified: in the Human Development Report 1994 (introducing the concept of human security) and in the 2001 report by the International Commission on Intervention and State Sovereignty (ICISS; introducing the concept of the Responsibility to Protect). In both instances, the climate action and atrocity prevention communities were brought together under a common concern about the protection of vulnerable human populations. But just as the space between climate impacts and human protection was shrinking in the atrocity prevention discussion, the space was beginning to widen in the climate science discussion.

The second section examines the period from the start of the so-called 'War on Terror' to the Arab Spring (2001–2011). It shows that the R2P policy development and institutionalisation process gradually marginalised and eventually removed climate-related concerns from its discussion. In addition, the Intergovernmental Panel on Climate Change (IPCC), drawing on qualitative scholarly literature (not quantitative field-based evidence) continued to insert more and more variables between climate impacts and conflict situations. Paradoxically, public concern about the threat posed by

DOI: 10.4324/9780367201364-4

climate change increased and the UN Security Council began what would become a series of debates on the implications of climate change on international peace and security. Ultimately, this was a period of flux that involved significant moments of divergence (including Cyclone Nargis in Myanmar) and few instances of convergence between the climate action and R2P agendas.

The third section explores the period from the Arab Spring to the post-pandemic world (2011–2022). It shows that the R2P and climate action agendas became increasingly embedded in the UN system in this period; however, in both cases, the 'pledges' made by governments to prevent atrocities and reduce emissions failed to materialise. It also flags that a new epoch of convergence between elements of the R2P agenda and the climate action agenda emerged and accelerated throughout this period. A discussion detailed in full in Chapter 3.

Background

The horrors of the Second World War and the emergent understanding of the extent of The Holocaust prompted a surge of activity under the generalised banner, 'never again!'. Described as 'the most significant international trial in history' (in part for establishing the legal precedent of individual criminal responsibility), in 1945, trials commenced in Nuremberg against leading Nazi figures for their role in organising and perpetrating 'war crimes', and for the first time in international law, 'crimes against humanity'.[1] The term 'genocide' did not exist in criminal law at this time, but was developed and defined during the final years of the Second World War by Raphael Lemkin, and after concerted advocacy, in 1948, the *Convention on the Prevention and Punishment of the Crime of Genocide* ('Genocide Convention') was adopted by the UN General Assembly, and subsequently ratified by 196 countries.[2] In the same year, the UN General Assembly adopted the Universal Declaration of Human Rights, enshrining a list of rights and freedoms that all humans should enjoy. And in 1949, the Third and Fourth Geneva Conventions were negotiated regarding the protection and treatment of prisoners-of-war and civilians in and around war zones, respectively.[3]

Global concern about the impact of industrialisation on human populations and the environment can be traced back to Rachel Carson's book, *Silent Spring* (1962), which identified the indiscriminate use of industrial chemicals, especially DDT, as a potential cause of illness and death.[4] A series of industrial accidents, such as a major oil tanker disaster off the English coast in 1967, further increased public concern in the West about industrial pollution.[5] In 1972, two publications, *Limits to Growth* and *A Blueprint for Survival*, began to link untrammelled industrial expansion with environmental degradation, global warming, and the potential decimation of the human population.[6] In the same year, the Stockholm Declaration – the main document generated by world's first UN conference on the environment – similarly warned of the 'incalculable harm to human beings' that may accompany the current misapplication of industrialisation, and delivered a broad brush Action Plan to correct the situation.[7] The second Villach conference on climate change, held in 1985, for the first time called on climate scientists and policymakers to work together to avoid by the year 2030, carbon dioxide concentrations in the atmosphere doubling, and global mean temperature rising by 2°C or thereabout.[8] In 1987, after a three-year process of research and consultation, the landmark report titled *Our Common Future* (or the Brundtland Report) was released.[9] Unlike the key idea underpinning Limits to Growth – that more industrial activity meant more environmental damage (including carbon emissions) that ultimately threatened humanity – Our Common Future argued that industrialisation could, and should, be *greened* (e.g. 'low emissions growth') in order to tackle the twin challenges of environmental damage and global inequality.[10] Similarly to its forebears however, the Brundtland Report warned that industrialisation left unchecked would cause significant environmental stresses that would likely trigger dire consequences for humanity including 'political and military conflict'[11] and, independently of the former, the 'mass movement of refugees'.[12]

These processes, ideas, and concerns would help inform and shape post-Cold War policymaking on emissions reductions and atrocity prevention.

Post-Cold War to the World Trade Center Attacks

In 1988, the first major international policy conference on global warming, titled *The World Conference on the Changing Atmosphere: Implications for Global Security*, was held in Toronto, Canada.[13] The event brought together more than 300 participants spanning policy and science (mostly from Western countries).[14] The Chair of the Brundtland Commission, Gro Harlem Brundtland, delivered the opening keynote, stating that 'many of the threats to the environment are truly global in scale and raise crucial questions of planetary survival'.[15] Reflecting these dire warnings, the opening sentence of the Toronto Conference Summary Statement – the key document to emerge from the conference – stated that 'Humanity is conducting an unintended, uncontrolled, globally pervasive experiment whose ultimate consequences could be second only to a global nuclear war'.[16] Based on these pronouncements, the chief policy recommendation in the Summary Statement was for industrialised countries to 'reduce carbon dioxide emissions by approximately 20 per cent of 1988 levels by the year 2005 as an initial global goal'.[17]

In 1990, the newly established IPCC released its First Assessment Report (AR1) on the impacts of climate change (Working Group 2).[18] The AR1 provided a synthesis of the global research on climate science and presented several possible future warming scenarios. On the latter issue, it wrote that 'the projected pace of global warming is unprecedented in human history (ten to fifty times faster compared to the earth's recovery from the last ice age, reaching 1.1°–5.5°C by 2100)', which 'could overwhelm humanity's ability to adapt, triggering widespread refugee problems, famine and conflict over scarce resources.'[19] The IPCC's supplementary report on impacts, released in 1992, would again identify links between climate impacts (droughts and floods especially) and large-scale human migration; as well as offer some basic adaptation solutions such as improved agricultural practices.[20]

Mounting concern about the impact of global warming on human and ecological systems motivated the UN General Assembly to establish the Intergovernmental Negotiating Committee (INC) for a Framework Convention on Climate Change in 1990. The INC's task was to formalise, through an international treaty, a framework for international cooperation to combat climate change by limiting global temperatures and coping with unavoidable climate impacts.[21] The INC held five sessions where more than 150 states discussed binding commitments, targets, and timetables for emissions reductions, financial mechanisms, technology transfer, and the 'common but differentiated' responsibilities of developed and developing countries.[22] In May 1992, the text of the United Nations Framework Convention on Climate Change (UNFCCC) was adopted at the United Nations Headquarters in New York. The landmark document centralised the belief that unmitigated climate change was dangerous to human and ecological systems, most prominently, stating under Article 2:

> The ultimate objective of this Convention (is the), stabilization of greenhouse gas concentrations in the atmosphere at a level that would prevent *dangerous* anthropogenic interference with the climate system [emphasis added].[23]

The UNFCCC secretariat later clarified that the UNFCCC text (and treaty) 'bound member states to act in the interests of human safety even in the face of scientific uncertainty', which offered some degree of clarification that 'dangerous' in the text referred to the threat that climate change posed to human populations.[24] However, the UNFCCC text stopped short of defining a level at which climate change may became dangerous both in terms of atmospheric concentrations of carbon dioxide and global average temperatures[25] – despite a general understanding among climate scientists that global temperatures exceeding 2°C above preindustrial levels would be unprecedented in human history and, by implication, unknown and dangerous.[26] In June 1992, the UNFCCC was signed by 154 countries at the United Nations Conference on Environment and Development, also known as the 'Earth Summit'. In March 1994, the UNFCCC treaty entered into force (and to date has been ratified by 197 countries).

In January 1992, with the state-based tensions of the Cold War firmly in the rear-view mirror, the UN Security Council met to discuss potential future threats to international peace and security and collective responses to them. After deliberations, the Security Council issued a statement explaining that it was 'time for a change' in thinking about threats and their solutions:

> The absence of war and military conflicts among States does not in itself ensure international peace and security. The non-military sources of instability in the economic, social, humanitarian and ecological fields have become threats to peace and security. The United Nations membership as a whole, working through the appropriate bodies, needs to give the highest priority to the solution of these matters.[27]

Based on a concern about how to deal with new non-military threats, the Security Council requested that the United Nations Secretary-General, Boutros Boutros-Ghali, prepare a suite of recommendations for future reforms to strengthen collective peacemaking and peacekeeping.[28] In June 1992, the Secretary-General delivered his pathbreaking report in response, titled *An Agenda for Peace: Preventive Diplomacy, Peacemaking and Peacekeeping*.[29] The report was primarily concerned with threats to peace beyond the ecological field; however, 'ecological damage' was identified as a new risk to global stability with pollution potentially posing a greater threat to a human population than 'a hostile army', while drought having the potential to 'decimate no less mercilessly than the weapons of war'.[30] With regard to solutions, the report proposed a framework spanning prevention, peacekeeping, peacemaking, and post-conflict peacebuilding – each populated with a suite of policy options closely resembling the R2P framework and policies.[31] The report emphasised the UN's role 'to try through diplomacy to remove the sources of danger before violence results', identifying preventive diplomacy as 'the most desirable and efficient' tool.[32] The report also identified strengthening early warning systems with regard to environmental threats and natural disasters and implementing water sharing programs in post-conflict societies as options that can help avoid threats to peace.[33]

In 1993, after mounting evidence of atrocities, the UN Security Council established the International Criminal Tribunal for the Former Yugoslavia (ICTY), referring to the reported programme of mass killings, mass rape, and the holding of territory, especially in Bosnia and Herzegovina, as 'ethnic cleansing'.[34] The idea of ethnic cleansing can be traced back to Antiquity – as can the notion of 'war crimes', 'crimes against humanity', and 'genocide' – but the Bosnian war gave this term global currency. In the end, the ICTY indicted 161 war criminals including the President of the former Yugoslavia, Slobodan Milošević, who was charged with genocide – and was the first head of state in human history to be brought to international legal accountability for crimes committed as a result of his rule.[35] However, he died before a verdict was reached. The same year, 1993, Bosnia and Herzegovina filed a complaint against Serbia (Former Republic of Yugoslavia) with the International Court of Justice (ICJ), claiming it had violated the Genocide Convention.[36] Thirteen years later, the ICJ found that while the mass killings were probably war crimes and crimes against humanity, they did not constitute genocide, apart from the Srebrenica massacre that was an 'act of genocide'.[37]

In 1994, the UN Development Programme would release its landmark *Human Development Report 1994*.[38] The central idea in the report, the concept of human security, directed attention towards the security of people and away from the security of nation-states.[39] This simple reframing had major ramifications. First, it began to challenge the traditional notion of state sovereignty (as we shall see, this was a core achievement of the R2P concept).[40] Second, it served to underscore the threats posed by environmental damage and climate change including – as explained in detail in the report – floods and cyclones, deforestation and habitat loss, droughts and water insecurity, sea level rise and saline penetration, and a variety of atmospheric pollutants including greenhouse gases (GHGs) that caused global warming.[41] In other words: the climate action and atrocity prevention agendas converged under the banner of human protection as emphasised by the concept of human security.

In 1995, the Parties to the UNFCCC met in Berlin at the First Conference of the Parties (COP1), presided over by German Environment Minister (and future German Chancellor) Angela Merkel. The Parties agreed that the commitments under the 1992 Convention were inadequate to meet its objective – to avoid 'dangerous' warming – and therefore must negotiate a more specific and quantifiable agreement than the Convention provided. The outcome document, the Berlin Mandate, established an ad hoc committee to negotiate a protocol by 1997 containing additional commitments for industrialised countries for the post-2000 period. After two years of negotiations, the Kyoto Protocol was adopted at the Third Conference of the Parties (COP3) held in Kyoto, Japan in 1997 – it would be the world's first GHG emissions reduction treaty.

The Kyoto Protocol operationalised the UNFCCC by committing industrialised countries to limit and reduce GHG emissions in accordance with agreed individual targets and the objective of the Convention. The Kyoto Protocol (1997) text states:

> In pursuit of the ultimate objective of the Convention as stated in its Article 2 [to avoid dangerous climate change] … Article 3 (1): The Parties included in Annex I [developed countries] shall, individually or jointly, ensure that their aggregate anthropogenic carbon dioxide equivalent emissions of the greenhouse gases listed in Annex A do not exceed their assigned amounts … with a view to reducing their overall emissions of such gases by at least 5 per cent below 1990 levels in the commitment period 2008 to 2012.[42]

The Protocol would legally bind developed country Parties (known collectively as Annex 1 Parties) to emission reduction targets over two commitment periods. In the first commitment period, 2008–2012, developed countries were required to cut their emissions of GHGs by an average of 5 per cent below 1990 levels, with specific targets varying from country to country. In the second commitment period, 2013–2020, the Parties would commit to reducing GHG emissions by at least 18 per cent below 1990 levels – as eventually agreed to under the 2012 Doha Amendment.[43] The Protocol also established three 'flexible mechanisms' to assist Annex I parties in meeting their national targets: Emissions Trading; Joint Implementation; and the Clean Development Mechanism. At their core, these mechanisms established the flow of carbon credits that ultimately disincentivised an industrial transition in some countries. The Kyoto Protocol also dropped direct reference to the word 'dangerous', and therefore could not identify a warming threshold, despite the European Council declaring in 1996 that 'global average temperatures should not exceed 2 degrees above pre-industrial level' – the first political body to formally identify what constituted dangerous warming.[44] In February 2005, the Kyoto Protocol entered into force, and by 2007 had been ratified by 192 countries.

In the late 1990s, the IPCC would release two more key reports that cautiously linked climate change to violent conflict. In 1995, the IPCC released its Second Assessment Report (AR2) on the impacts of climate change, human vulnerability, and adaption.[45] The report found that environmental degradation and associated resource allocation 'may be potent factors in provoking conflict', and 'by affecting water and food supply, climate change could increase the possibility of violent conflicts in a number of regions.'[46] This report, and the follow up IPCC Special Report on climate impacts of 1998, identified a suite of policy options 'to avoid conflict' over water resources such as joint legal agreements, pricing systems, efficiency initiatives, infrastructure improvements, agricultural and urban planning policies, among many others.[47] These could be considered direct human protection policies in contrast to the indirect human protection policies pursued via emissions reductions and avoided warming.

The 1994 Rwandan Genocide and the 1995 Srebrenica massacre (see Chapter 1) focused the relevant global policymaking communities on coming up with new ideas and approaches to atrocity prevention and response. By 1996, Francis Deng and his Brookings colleagues had developed the concept of 'sovereignty as responsibility', which argued that 'sovereignty can no longer be seen as a protection against interference, but as a charge of responsibility where the state is accountable to both domestic and external constituencies'.[48] Put simply, Deng believed that while state sovereignty was important, 'these days' states also have 'positive responsibilities to protect their citizens'.[49] In 1998, Deng and Roberta Cohen[50] identified several possible consequences should states fail to protect their populations: 'States that refuse access to populations at risk could expect calibrated actions ranging from diplomacy to political pressure, sanctions, and, as a last resort, military intervention'.[51] Deng's and Cohen's extensive professional experience with Internally Displaced Persons drove the development of the concept; yet despite this, climate-related issues were mentioned only sparingly (with 'global warming' or 'climate change' not mentioned at all). The focus, when discussed, was on the relationship between natural resource allocation (especially water) and conflict, Deng arguing that unless resource sharing policies are improved in some regions, conflicts 'have a strong potential for escalating'.[52] He also repeatedly cited the important dual roles that the Intergovernmental Authority on Drought and Development play across North Africa in mediating civil conflicts as well as facilitating drought relief arrangements.[53] Deng's contribution, above all however, significantly shifted the international conversation on atrocity prevention from 'sovereignty as a right' to 'sovereignty as a responsibility' (adding to Human Security's individual protection discourse, and paving the way for R2P).

Also, in 1996, drawing heavily on the precedents of the ICTY (and Nuremberg) such as individual criminal responsibility, the UN Security Council created the International Criminal Tribunal for Rwanda (ICTR).[54] While criticisms abound, especially regarding delays in the prosecutions and the US$2 billion price tag, by 2015, after 21 years, the ICTR had sentenced more than 60 ringleaders of the genocide.

Two more major contributions to the sovereignty-intervention debate came in 1999. Against the backdrop of ongoing mass atrocities in Kosovo and the use of NATO airpower in response, in April, UK Prime Minister Tony Blair delivered a speech titled the Doctrine of the International Community. The Blair Doctrine, as it was subsequently called, emphasised the globalised nature of threats (rather than bounded threats), the importance of international cooperation between states (rather than UN-based cooperation) and prioritised military intervention (rather than prevention, less extreme coercive options or post-conflict peacebuilding).[55] These themes would underpin Blair's core message: that NATO bombardment in Kosovo was justified in response to the 'evil of ethnic cleansing'.[56] Blair's contribution rallied Western countries behind the virtues of military intervention for human protection purposes, but failed to resonate in the Global South or with the UN – in large measure, conflicting with the Agenda for Peace of 1992. Blair – showing consistency with this world view – mentioned that achieving the Kyoto targets 'to slow or stop global warming' was a matter for inter-state relations 'between the main industrial nations and the developing world'.[57]

In September 1999, with atrocities underway in East Timor (see Chapter 1), the UN Secretary-General Kofi Annan (replacing Boutros Boutros-Ghali in 1997) sought to push back against Blair's formulation. Annan argued that when mass atrocities are imminent or occurring, and the state in charge 'is unable or unwilling to stop it', the international community, acting via the authority of the UN Security Council, must weigh national sovereignty considerations against the 'individual sovereignty' of those in peril.[58] The idea of two sovereignties was based on Annan's observation that state sovereignty was being 'redefined' (perhaps seconded) by a variety of international processes that centralised individuals including economic, political, and social globalisation, and a growing commitment to human rights. Annan was emphatic that intervention to protect civilians 'should not be understood as referring only to the use of force' but rather encompass a commitment to prevention and peacebuilding activities.[59] Annan's idea intensified debate over sovereignty and intervention.

Soon after, in March 2000, the Secretary-General delivered a report titled *'We the Peoples': The Role of the United Nations in the 21st Century*. The report aimed to carry forward into the new Millennium the momentum from the previous decade on the changing nature of threats and appropriate responses to them. The report comprised five substantive chapters covering global governance, inequality and health, atrocity prevention, environment and climate change, and UN reform. Annan's chapter on atrocity prevention identified two key shifts that bridged the atrocity and climate discourses. First, major threats to human populations had shifted from interstate wars (during the Cold War) to intrastate mass atrocity situations (in the 1990s) requiring 'a more human centred approach to security' into the future.[60] Second, Annan wrote that while mass atrocity situations have to date been caused by 'greed and political ambition stoking ethnic and religious differences',

> As we look ahead, we can see real risks that resource depletion, especially freshwater scarcities, as well as severe forms of environmental degradation, may increase social and political tensions in unpredictable but potentially dangerous ways.[61]

As with the Human Security Report 1994, Annan's focus on human protection brought into sharper focus the threat of ecological damage. The report's chapter on climate change doubled down on this threat, repeatedly describing unsustainable human activities as threatening 'our children's future', 'the survival of our species', and 'the planet'.[62] Annan's list of possible solutions covered technology, markets, policy, and governance.

The remainder of the atrocity prevention chapter however was devoted to traditional solutions (excluding climate-related issues). Annan traversed nearly the full list of policy options that would ultimately be included in the R2P toolbox: from preventive diplomacy and political sanctions, through to legal strategies and promoting human rights, broad brush and targeted financial sanctions, arms embargoes, and emergency aid and peace operations. Regarding the 'dilemma of [military] intervention', Annan stopped short of argumentation, rather posing an open-ended question for the international community to consider:

> If humanitarian intervention is, indeed, an unacceptable assault on sovereignty, how should we respond to a Rwanda, to a Srebrenica – to gross and systematic violations of human rights that offend every precept of our common humanity?[63]

In September 2000, Canada's Foreign Minister, Lloyd Axworthy, responded to Annan's challenge by launching the *International Commission on Intervention and State Sovereignty* (ICISS). Co-chaired by Gareth Evans, the ICISS was tasked with providing the international community, via the UN, with an innovate approach to human protection that reconciled state sovereignty and intervention. After more than 12 months of consultation, in December 2001, the ICISS published its report under the title *The Responsibility to Protect*.[64]

The report made three main policy contributions. First, the authors recast sovereignty and intervention as responsibilities – both previously thought of as rights under the Westphalian tradition and the 'right to intervene' language, respectively.[65] It followed that 'the primary responsibility for the protection of its people lies with the state itself'. However, 'where a population is suffering serious harm … and the state in question is unwilling or unable to halt or avert it', or is indeed itself the perpetrator, 'the principle of non-intervention yields to the international responsibility to protect'.[66] Second, as might be required, the commissioners not only developed a comprehensive operational response framework that prioritised prevention, but also developed reaction and rebuilding phases – all phases displaying the full range of political, economic, legal, and security strategies and policies on offer (see Part 2 of the present book).[67] Third, the report identified the conditions that would justify 'military intervention for human protection purposes', namely, in order to halt or avert: actions defined under the Genocide Convention, different manifestation of ethnic cleansing, crimes against humanity and war-related crimes as defined in the Geneva Conventions, mass starvation as a consequence of state collapse and or civil war, and in response to 'overwhelming natural or environmental catastrophes, where the state concerned is either unwilling or unable to cope, or call for assistance, and significant loss of life is occurring or threatened'.[68] The ICISS report (perhaps inadvertently) presented the closest relationship in history between climate impacts and atrocity prevention policies. While early language around the planet destroying potential of global warming was shocking (above), nowhere in the mainstream climate policy discussion was climate-related military action proposed.

In 2001, the IPCC released its Third Assessment Report (AR3) on the impacts of climate change, human vulnerability, and adaption. The AR3 delivered more detail on the relationship between climate-related impacts and violent conflict compared to the AR2, but with it a strengthened reluctance to link the two. To be sure, in general, the AR3 contended, a causal connection between climate-related resource scarcity, human migration, and military conflict 'is possible', but the latest global research on this topic, it explained, has found that the 'many intervening and contributing causes' of violent conflict largely negate climate change as a key driver.[69] The AR3 delivered a comprehensive suite of policies across a range of climate-impacts to assist human populations adapt to unavoidable warming.

GHG mitigation policy was also progressing. In November 2001, the Seventh Conference of the Parties (COP7) was held in Marrakesh, Morocco. The chief outcome document, the Marrakesh Accords, consisted of a package of decisions on the operational rules of the Kyoto Protocol, including the 'flexible mechanisms' (ETS, JI, CDM), as well as reporting, compliance, and the treatment of forestry, which set the stage for ratification of the Kyoto Protocol.[70] This was a welcome relief after a dramatic failure to agree on the policy detail the previous year at COP6 held at The Hague, which was suspended until July 2001 and relocated to Bonn.[71] The Marrakesh Accords also addressed issues such as capacity building, technology transfer, responding to the adverse effects of climate change, and the establishment of three funds to assist vulnerable populations: the Least Developed Countries Fund, Special Climate Change Fund, and Adaptation Fund.

The War on Terror to the Arab Spring

By December 2000, George W. Bush had proven victorious over Al Gore in the US general election, and, in the aftermath of the September 2001 attacks, had reoriented US external (and internal) policy towards rooting out "Islamic terrorists" worldwide. The ICISS report, published in December 2001, was in danger of being shelved in perpetuity as fighting global terror took precedence over preventing internal human protection catastrophes. But momentum slowly built towards institutional acceptance of the R2P concept, and with each milestone came increased marginalisation of environmental threats.

In 2003, the UN Secretary-General established the High-level Panel on Threats, Challenges, and Change, to assess current and future threats to international peace and security and recommend ways in which the UN can help provide collective security in the decades to come.[72] In December 2004, the High-level Panel delivered its final report titled *A More Secure World: Our Shared Responsibility*.[73] The report – and first R2P milestone – identified 'six clusters of threats' (reflected as chapters) covering terrorism, transnational crime, weapons of mass destruction, interstate conflict, and 'internal conflict, including civil war, genocide and other large-scale atrocities', and bundled together, 'economic and social threats, including poverty, infectious disease and environmental degradation'.[74] When compared to Annan's Millennium report of 2000 (above), which devoted a full chapter to the environment and climate, the language in the 2004 report to frame environmental threats was significantly subdued (e.g. 'planetary destruction' verses 'environmental concerns').[75] This downsizing of the climate threat filtered into the report's R2P discussion, with the crucial recommendation on R2P expressed in the following terms:

> We endorse the emerging norm that there is a collective international responsibility to protect, exercisable by the Security Council authorizing military intervention as a last resort, in the event of genocide and other large-scale killing, ethnic cleansing or serious violations of international humanitarian law which sovereign Governments have proved powerless or unwilling to prevent.[76]

This update on the conditions that would justify military intervention for human protection purposes excluded 'overwhelming natural or environmental catastrophes' as expressed in the ICISS report of 2001 (above). However, the report was highly critical of industrialised countries, and emerging economies to a lesser extent, for failing to reduce their 'excessive dependency on fossil fuels' and collectively failing to implement the Kyoto Protocol's 5 per cent emission reduction target and negotiating a post-2012 agreement. It recommended that countries begin to reduce subsidies on fossil fuels, increase incentives for renewable energy, and take immediate meaningful action to remedy both failings of the Kyoto Protocol process.[77]

Aside from the R2P policy development process, but related nonetheless to human protection policy development in this period, in 2003, the Extraordinary Chambers in the Courts of Cambodia (ECCC) was established to try the key figures responsible for the genocidal crimes committed in Cambodia in the late 1970s (see Chapter 1).[78] Unlike the purely international ICTY or ICTR (above), the ECCC combined international and domestic legal authorities, which was primarily due to the UN and Cambodia not being confident in each other's legal processes.[79] The hybrid process was plagued with problems throughout, including, similarly to the ICTY and ICTR, running hugely over budget and the accused regularly dying of old age before standing trial.[80] After 16 years, only three Khmer Rouge officials were convicted, producing little by way of justice for the past or deterrence for the future.[81]

The second major milestone towards formal acceptance of the R2P concept came in March 2005, with the Secretary-General's report on UN reform titled *In Larger Freedom: Towards Development, Security and Human Rights for All*.[82] For the R2P to progress, it was essential that the key recommendations expressed in *A More Secure World* (2004) were included in the potential reforms expressed in *In Larger Freedom* (2005).[83] Success came with the Secretary-General's reform blueprint formally urging heads of states and government to

> Embrace the 'responsibility to protect' as a basis for collective action against genocide, ethnic cleansing and crimes against humanity, and agree to act on this responsibility, recognizing that this responsibility lies first and foremost with each individual State, whose duty it is to protect its population, but that if national authorities are unwilling or unable to protect their citizens, then the responsibility shifts to the international community to use diplomatic, humanitarian and other methods to help protect civilian populations, and that if such methods appear insufficient the Security Council may out of necessity decide to take action under the Charter, including enforcement action, if so required.[84]

The Secretary-General's report had hit all the right notes largely reflecting not only the 2004 recommendations but also the original ICISS/R2P report of 2001. However, the blueprint report also served to consolidate the exclusion of climate-related human catastrophes as a justification for intervention under the R2P. Indeed, environmental and climate considerations in the 2005 blueprint were scarce, with threat framing and mitigation policy solutions consistent with the 2004 report.[85] It did however recommend the 'establishment of a worldwide early warning system' to protect populations from 'all natural disasters' including storms, floods, droughts, landslides, and heat waves.[86]

In September 2005, more than 150 heads of state and government gathered in New York for the World Summit to commemorate the 60th anniversary of the UN, and to conclude negotiations on Annan's blueprint for UN reform.[87] The World Summit Outcome Document, which was unanimously adopted by all 193 Member States of the UN, set out the international community's commitment to UN reform and represented the formal institutional acceptance of the R2P that had been four years in the making. The Outcome Document as related to R2P was titled 'Responsibility to protect populations from genocide, war crimes, ethnic cleansing and crimes against humanity' under which the text read in full:

> 138. Each individual State has the responsibility to protect its populations from genocide, war crimes, ethnic cleansing and crimes against humanity. This responsibility entails the prevention of such crimes, including their incitement, through appropriate and necessary means. We accept that responsibility and will act in accordance with it. The international community should, as appropriate, encourage and help States to exercise this responsibility and support the United Nations in establishing an early warning capability.
>
> 139. The international community, through the United Nations, also has the responsibility to use appropriate diplomatic, humanitarian and other peaceful means, in accordance with Chapters VI and VIII of the Charter, to help to protect populations from genocide, war crimes, ethnic cleansing and crimes against humanity. In this context, we are prepared to take collective action, in a timely and decisive manner, through the Security Council, in accordance with the Charter, including Chapter VII, on a case-by-case basis and in cooperation with relevant regional organizations as appropriate, should peaceful means be inadequate and national authorities are manifestly failing to protect their populations from genocide, war crimes, ethnic cleansing and crimes against humanity. We stress the need for the General Assembly to continue consideration of the responsibility to protect populations from genocide, war crimes, ethnic cleansing and crimes against humanity and its implications, bearing in mind the principles of the Charter and international law. We also intend to commit ourselves, as necessary and appropriate, to helping States build capacity to protect their populations from genocide, war crimes, ethnic cleansing and crimes against humanity and to assisting those which are under stress before crises and conflicts break out.
>
> 140. We fully support the mission of the Special Adviser of the Secretary-General on the Prevention of Genocide.[88]

The Outcome Document (September 2005) also confirmed beyond doubt that the R2P *did not* apply to climate-related human catastrophes. This was consistent with the key lead up documents (December 2004, March 2005), but inconsistent with the original ICISS report (December 2001). While consistency in the reports immediately preceding the Summit would suggest a fait accompli on the R2P scope in the Outcome Document, those involved in the negotiations explain that final agreement was only settled after months of in-house negotiations preceding the 2005 Summit.[89] According to Edward Luck, the UN Secretary-General's Special Adviser on the Responsibility to Protect from 2008 to 2012, Member States were nervous that 'the ICISS report had really been rather broad on this score, not defining the scope of R2P crimes in a specific or consistent manner', especially as it had included environmental emergencies, and therefore they 'wanted to delineate the scope of R2P's application as precisely as possible'.[90] Luck explained that confirming R2P's four-fold scope (genocide, crimes against humanity, war crimes, and ethnic cleansing) delivered a marked transformation in the final month of negotiations from many Members who were 'decidedly negative' about R2P to at the World Summit being able to 'agree on rather detailed consensus language on R2P' (this discussion is expanded upon in Chapter 3).[91]

Climate change was specifically addressed elsewhere in the Outcome Document, and consistent with lead up reports, was framed as 'serious' and a 'concern', with GHG mitigation options spanning policy, finance, and technology. Adaptation policy was given a considerable boost compared to previous reports, emphasising policies across water, desertification, biodiversity, natural disasters, forests, and energy to help protect vulnerable communities.[92] The Outcome Document was subsequently unanimously adopted by 193 Member States of the UN General Assembly. In April 2006, after six more months of negotiations, the UN Security Council unanimously reaffirmed the World Summit's conclusion related to R2P, its prevention focus, and the four crimes, in a thematic resolution (1674) on the protection of civilians in armed conflict.[93] Institutional acceptance had been achieved.

Efforts to curb global warming also took a considerable leap in 2005. In January, the European Union Emissions Trading Scheme (EU ETS) was launched, the first and still largest cap and trade system in the world. The EU ETS required major emitting facilities – collectively responsible for close to half of the EU's total carbon dioxide emissions – to start to pay for the carbon emissions that they produced. However, dramatic carbon price fluctuations over the ensuing years (e.g. from €30 a tonne in April 2006 to €0.10 in September 2007) would do little to pressure firms to change their production processes in the immediate term (the price reached €99 a tonne in 2022). In February, the Kyoto Protocol entered into force – becoming a legally binding document – after Russia submitted its instrument of ratification.[94] Under the terms agreed in Kyoto, the Protocol would only enter into force following ratification by 55 Parties to the UNFCCC, and if those 55 countries included enough Annex I (industrialised) Parties that at least 55 per cent of that group's total carbon emissions were represented. Russia's actions satisfied these prerequisites. However, disappointment was not too far away with the world's largest carbon emitter, the US, refusing to ratify the treaty under the leadership of President George W. Bush.

By 2007, public concern about the security implication of climate change had increased,[95] manifesting in heightened activity inside and outside the UN. For example, in April 2007, the Security Council discussed for the first time the linkages between climate and security, the UN Security Council President opening debate in the following way:[96]

> Our responsibility in the Council is to maintain international peace and security, including the prevention of conflict. An unstable climate will exacerbate some of the core drivers of conflict, such as migratory pressures and competition for resources.[97]

In May, the US-based Center for Naval Analysis released a highly influential report titled *National Security and the Threat of Climate Change*.[98] Its authors, 11 retired US military officials, presented a similar formulation to the UN Security Council, as expressed in the report's foreword: 'Climate change can act as a threat multiplier for instability in some of the most volatile regions of the world, and it presents significant national security challenges for the United States'.[99] This was the first articulation of climate change as a 'threat multiplier' – serving to exacerbate or amplify existing political, social, economic, and/or environmental threats and tensions (for an expanded discussion see Chapter 3). However, in contrast to the surge of interest in the climate-conflict relationship, the IPCC Fourth Assessment Report (AR4) on impacts and vulnerability, also released in 2007, provided only limited discussion with very few references.[100] Subsequent analysis of the AR4 found that it offered 'little or no support for a climate change to armed conflict scenario' including on water stress, which had been central to the climate-conflict discussion in previous IPCC reports.[101]

In January 2007, Ban Ki-moon – a strong supporter of the climate action agenda – replaced Kofi Annan as UN Secretary-General. Opening the proceedings at the Thirteenth Conference of the Parties (COP13) held in Bali, Indonesia, in December 2007, Secretary-General Ki-moon framed climate change as the 'defining challenge of our time' and urged immediate action to reduce carbon emissions while promoting economic growth (a formula that the Brundtland Report had centralised in global climate negotiations 20 years prior).[102] Building on the repeated recommendations from a string of major UN reports such as *A More Secure World* (December 2004, p 30), *In Larger Freedom* (March 2005 p 20), and the World Summit *Outcome Document* (September 2005 p 12), the Bali COP was to, above all, produce an 'action plan' for the negotiation of a successor to the Kyoto Protocol, due to expire in 2012. Before agreeing on an action plan however, delegates were hoping to confirm whether industrialised countries would agree to adopt a strengthened emissions reduction target in the post-Kyoto period: up from 5 per cent by 2012 to between 25 per cent and 40 per cent by 2020 as recommended by the IPCC AR4 and subsequently broadly accepted at an international climate meeting in Vienna in August 2007.[103] The Bali negotiations produced the Bali Action Plan that similarly to previous COPs 'recognized that deep cuts in global emissions will be required to achieve the ultimate objective of the Convention [to avoid warming that is dangerous to human populations]'.[104] However, the term 'dangerous' did not feature explicitly, nor did the temperature increase that might be considered dangerous.[105] In addition, reportedly at the insistence of the US delegation, the Plan excluded any reference to strengthened emission reduction targets for industrialised countries, and instead emphasised the need for developing countries to adopt binding emissions commitments.[106]

Between 2003 and 2010, the practical application of the R2P would have some successes and some failures. The first test came almost immediately after the ICISS report of 2001, in responding to the government of Sudan and their 'Janjaweed' allies conducing a scorched earth campaign in the Darfur region in 2002–2003, resulting in the death of approximately 300,000 people.[107] While clearly an R2P situation that required a robust, perhaps military, response to protect civilians, for a variety of reasons such as conceptual underdevelopment and/or the failure of R2P advocates, intervention did not occur. This disappointing start continued when in August 2006 the UN Security Council adopted Resolution 1706 on the situation in Darfur. The Resolution invoked the R2P concept to mandate an extension to the UN Mission in Sudan peacekeeping operation.[108] For some scholars, Resolution 1706 has 'gone down in history as one of the Council's worst ever' because the mandated mission was unlikely to happen, and the ferocity of the internal debate that proceeded it discouraged the Council from revisiting the R2P in the context of Darfur.[109] In contrast, the R2P was invoked with great success in response to the highly volatile situation unfolding in Kenya in late 2007 where claims of election rigging were descending into ethnic and tribal violence. With more than 1,000 people killed and 300,000 displaced, the African Union appointed Kofi Annan to lead a mediation team in the hope of negotiating a political settlement. Annan approached the situation, as he put it, 'in the R2P prism', which was echoed by Secretary-General Ki-moon, and his newly appointed Special Adviser on the Prevention of Genocide and Mass Atrocities, Francis Deng (above).[110] Annan successfully persuaded the country's political leaders to reach a power sharing agreement and rein in their supporters. Some scholars describe this situation as providing 'a tangible demonstration of R2P's capacity to facilitate atrocity prevention through peaceful means'.[111]

In 2008, several highly controversial human protection situations emerged that would test the operational limits of the R2P. In May 2008, Cyclone Nargis with its massive tidal surge devastated Myanmar's Irrawaddy Delta, directly killing 130,000 people. The Myanmar government's shockingly slow response to accept international humanitarian assistance – and with thousands of lives assumed at risk from disease, starvation, and exposure as a result – caused the French Foreign Minister, Bernard Kouchner, to argue that this was an R2P situation that justified intervention – at least in the form of air dropping or barge landing relief supplies – without the government's consent.[112] The proposal attracted little support, primarily on the basis that R2P did not justify coercive military intervention in response to natural disasters, only in response to the four crimes enshrined in the World Summit Outcome Document.[113] A few months later, in August, Russia justified its military invasion of Georgia on R2P grounds – to halt what it claimed was a genocide occurring there. Unlike Kouchner's appeal to an expanded R2P scope, the central controversy in this instance was that R2P did not permit the unilateral use of force; and secondly, there was no evidence of genocide. And in the final months of 2008, the Sri Lankan government launched a military campaign to annihilate the Tamil Tigers – labelled a terrorist organisation by many countries. Up to 40,000 civilians were killed, the great majority at the hands of government forces.[114] The Security Council did not even bother to meet on the unfolding crisis, despite the 'conscience-shocking' nature of the atrocities, in effect sidelining the R2P concept.

These controversies inflamed general anxieties about R2P's potential misuse. Gareth Evans recalls that throughout 2008, 'the language of "buyer's remorse" was in the air', with some African, Latin American, and Arab countries even rejecting the proposition that the World Summit had endorsed the R2P *concept*; rather it had simply endorsed the view that civilians should be protected from the four crimes – a view that Evans later described as 'nonsense on stilts'.[115] Edward Luck, who by 2008 had been appointed the Secretary-General's Special Adviser on the R2P, assessed the 2008 situation as a general concern by UN delegates that the scope of the R2P might be extended beyond the four crimes, subsequently writing that

> When we were trying to get the member states back on board in 2008, some of them would come to see me rather frequently to get reassurance that the Secretary-General's report I was drafting [below] would stay within the bounds of the four specified crimes and would not endorse wider application of R2P. It seems that going even an inch beyond would open up for some of them a Pandora's Box, as well as increasing the opportunities for other states to misapply the principle. Member states become distinctly uncomfortable when confronted by what they perceive to be a boundless concept.[116]

In January 2009, the Secretary-General delivered his report titled *Implementing the Responsibility to Protect*.[117] It reaffirmed the World Summit's key components relating to the R2P and argued that the best way to discourage R2P's misuse was to develop its standards, tools, strategies, and policies – in essence, more R2P was what was needed, not less.[118] The Secretary-General also confirmed that the scope was fixed, for now:

> The responsibility to protect applies, until Member States decide otherwise, only to the four specified crimes and violations: genocide, war crimes, ethnic cleansing and crimes against humanity. To try to extend it to cover other calamities, such as HIV/AIDS, climate change or the response to natural

disasters, would undermine the 2005 consensus and stretch the concept beyond recognition or operational utility.[119]

In September 2009, the UN General Assembly passed Resolution 63/308 recognising the Secretary-General's report.[120] And in November, the UN Security Council adopted Resolution 1894, which reaffirmed the 2005 Outcome Document as related to R2P including 'the responsibility to protect populations from genocide, war crimes, ethnic cleansing and crimes against humanity'.[121]

Throughout 2009, the Secretary-General was also closely engaged in the emerging climate security discussion playing out in the UN. In June, the UN General Assembly passed Resolution 63/281, proposed by Pacific Island states vulnerable to climate impacts,[122] requesting that the Secretary-General prepare a report 'on the possible security implications of climate change'.[123] The report (A/64/350), submitted to the General Assembly in September, delivered a two-pronged message that would (probably inadvertently) encompass features of the R2P agenda. First, the Secretary-General's report would define climate change as a 'threat multiplier' – that is, it exacerbated existing threats to peace and security including 'persistent poverty, weak institutions for resource management and conflict resolution, fault lines and a history of mistrust between communities and nations, and inadequate access to information or resources'.[124] Ki-moon's list of threats were consistent with the ICISS report's 'root causes' of deadly conflict.[125] Second, the Secretary-General identified a series of 'threat minimisers' – that is, policy actions that were capable of diminishing the aforementioned threats, which included those related to 'economic development, democratic governance and strong local and national institutions, international cooperation, preventive diplomacy and mediation'.[126] Ki-moon's solutions list closely resembled the R2P's atrocity prevention strategies and policies.[127]

The Fifteenth Conference of the Parties (COP15) was held in Copenhagen in December 2009. The much-anticipated COP was the largest international climate meeting in history, involving nearly 27,000 people, including 10,500 delegates representing 190 states, and over 120 heads of state and government.[128] The main goal was to establish a legally binding agreement for global emissions reductions for the Kyoto Protocol's second commitment period (2013–2020). On 19 December, the Copenhagen Accord was reached through in series of back-room deals between the US (with President Barak Obama now in The White House), China, India, Brazil, and South Africa. The Accord, in a first for COP events, described rising temperatures as dangerous to human populations *and* identified a temperature goal that should not be breached, the signatories agreeing,

> To achieve the ultimate objective of the Convention to stabilize greenhouse gas concentration in the atmosphere at a level that would prevent dangerous anthropogenic interference with the climate system, we shall, recognizing

the scientific view that the increase in global temperature should be below 2 degrees Celsius.[129]

The Accord also provided a framework for states to register new national emission reduction targets called 'pledges'. The pledges, which were not legally binding, avoided developed and developing countries alike committing to firm targets for the next period of the Kyoto Protocol. By April 2010, 120 Parties, together responsible for about 80 per cent of global emissions, had nominated mitigation pledges for the period up to 2020. Collectively however the pledges would not avoid dangerous warming, but rather amount to a projected temperature increase of about 4°C above pre-industrial levels by 2100.[130]

The Arab Spring to the Post-Pandemic World

In December 2010, a Tunisian street vendor set himself on fire in protest at repeated harassment by state authorities.[131] This single act triggered the 'Arab Spring' – a popular uprising against oppressive regimes in the Arab world. In early February 2011, the Arab Spring reached Libya. Initial protests in Benghazi quickly turned into a countrywide uprising that threatened to topple the country's dictator, Muammar Gaddafi, who had been in power for more than 40 years.[132] Gaddafi's forces responded by repeatedly deploying lethal force against protesters, killing several hundred throughout February, and went door-to-door arresting and disappearing suspected agitators.[133] By late February, the government had lost control of towns and cities to the east of the country. As Gaddafi's forces continued to murder civilians in great numbers, Gaddafi himself began to broadcast dehumanising threats reminiscent of the Rwandan genocide, referring to protesters as 'cockroaches' and 'rats', and pledging to 'cleanse Libya house-by-house'.[134]

In response, the UN Secretary-General Ban Ki-moon communicated his assessment to the UN Security Council, calling the situation 'totally unacceptable' and emphasising the Libyan government's 'responsibility to protect it people'.[135] Several heads of UN departments, including R2P leaders Ed Luck and Francis Deng, and relevant regional organisations such as the African Union and Arab League, strongly condemned Gaddafi's behaviour and what appeared to be his incitement to commit grave human rights violations. On 26 February, the UN Security Council adopted Resolution 1970 that referred to the Libyan authorities' responsibility to protect its population and issued a range of non-military measures to halt the violence, including, diplomatic peacemaking, arms embargoes, asset freezes, and referral to the International Criminal Court.[136] Subsequent attempts to negotiate a political settlement repeatedly failed, and in March the Arab League called for a no-fly zone over Libya. All these efforts did not persuade Gaddafi's forces to stop killing civilians, and as they reached the outskirts of Benghazi there was wide anticipation in Western capitals, in regional organisations, and among the UN leadership that a large-scale massacre of civilians was imminent. On 17 March, the UN Security

Council passed Resolution 1973 that authorised 'all necessary measures' including military action and a no-fly zone to protect Libyan civilians.[137] This was the first time in Council history that it had authorised the use of force for human protection purposes without the consent of the recognised government concerned.[138] Two days later, a NATO-led alliance, initiated by the US, UK, and France with the support of 18 other states including several Arab counties, conducted a massive aerial bombardment of Gaddafi's air defences and military forces positioned on the outskirts of Benghazi. The bombing campaign continued into late April, significantly diminishing government capability, and by May military momentum had shifted towards the anti-Gaddafi rebels. On 21 August 2011, the rebels captured Tripoli, Libya's capital. Gaddafi was subsequently found hiding in a drain and shot and stabbed to death in the street.

Efforts to rebuild Libya in the wake of the mayhem were later assessed to be under-resourced and unenergetically pursued via the UN Support Mission in Libya (UNSMIL), which contributed to an intensification of regional, tribal, and personal enmities, the rise of Islamic State in Libya, and a lengthy period of mass migration to southern Europe via the Mediterranean – all of which resulted in countless deaths.[139]

As could be expected, NATO's response in Libya attracted significant controversy. On the one hand, some argued that it was a high point for the R2P concept because it performed as it should and protected civilians from atrocity crimes. For example, R2P academic Alex Bellamy wrote that the concept 'played an important role in shaping the world's response to actual and threatened atrocities' in Libya, while Ramesh Thakur, also a leading R2P academic, explained that R2P was a 'game-changer' that acted 'as a powerful new galvanising norm' over Libya.[140] Conveying similar sentiments, the UN Secretary-General Ban Ki-moon explained that 'Our debates are now about how, not whether, to implement the responsibility to protect'.[141] R2P co-creator, Gareth Evans, delivered the strongest defence of the R2P-inspired intervention, stating that it 'unquestionably worked – certainly in preventing a major massacre in Benghazi', and represented a clear demonstration that R2P advocates had 'won the debate over stopping genocide'.[142] On the other hand, detractors, led by the BRICS countries (Brazil, Russia, India, China, and South Africa), argued that NATO, and the US, UK, and France, had blatantly used what was a mandated civilian protection operation for the purpose of 'regime change'.[143] India, for example, referred to NATO as a de facto air force for the rebels, rather than a defensive shield against populations being massacred, while South Africa asserted that 'we strongly believe that the resolution is being abused for regime change, political assassinations and foreign military occupation'.[144] In November 2011, even former UN Secretary-General, Kofi Annan, conveyed his disappointment that 'regime change came up very quickly' in Libya.[145]

Upon reflection, many supporters of the intervention in Libya conceded that so-called 'mission creep' (from human protection to regime change) did occur. For example, Evans wrote in 2016 that 'Early on, the NATO coalition decided to pursue not just the "protection of civilians and civilian-populated areas", as the UN Security Council mandate expressly provided, but the overthrow of the Qaddafi regime'.[146] Considerable evidence points in this direction; for example, in April 2011, during the NATO bombing campaign, US President Barack Obama, UK Prime Minister David Cameron, and President of France Nicholas Sarkozy wrote in joint opinion editorial:

> Our duty and our mandate under UN Security Council Resolution 1973 is to protect civilians, and we are doing that. It is not to remove Gaddafi by force. But it is impossible to imagine a future for Libya with Gaddafi in power.[147]

The Libyan intervention, and especially NATO's obscured position on the end game, had several significant consequences for the R2P conceptually, institutionally, and operationally. First, in late 2011, the Brazilian government published the idea of the 'responsibility while protecting' (RwP), which sought to clarify the criteria and oversight needed for UN Security Council mandates that authorise the use of military force.[148] Second, R2P became increasingly central to UN Security Council discussions. For example, between 2005 and 2009, only two UN Security Council resolutions referenced R2P in country situations (Darfur and DRC). By contrast, between 2010 and 2014, the UN Security Council passed 18 resolutions 'recognising', 'recalling', 'affirming', 'reaffirming', 'reiterating', 'underscoring' the responsibility of a variety of governments (Libya, Central African Republic, South Sudan, Syria, Cote d'Ivoire, Yemen, Mali, Somalia) to protect their populations.[149] Third, Libya's mission creep situation 'led to the BRICS refusing to take any action at all in response to the unfolding horror in Syria (in 2011), which in its early stages bore an uncanny resemblance to Qaddafi's oppression'.[150] As of 2021, the Syrian conflict has resulted in more than 400,000 deaths including, as UNICEF report, at least 'one child injured or killed every eight hours over the past 10 years'.[151]

In 2010, the Sixteenth Conference of the Parties (COP16) to the UNFCCC was held in Cancun, Mexico. In his opening address, the UN Secretary-General warned the delegates what was at stake and emphasised the need for enhanced climate action:

> We are here for one reason: to protect people and the planet from uncontrolled climate change. To do that, we need to make progress – in these global negotiations and through national actions each of you take in your countries to curb emissions and increase resilience.[152]

Based on the most up-to-date climate science, the delegates agreed through the Cancun Agreement that the international community should consider holding global warming to the safer temperature rise of 1.5°C above pre-industrial levels rather than 2°C.[153] However, two features of the Agreement would serve to undermine this strengthened temperature goal, namely, the establishment of carbon capture and storage (still commercially unproven) as a CDM program, and new deforestation guidelines – both of which presented opportunities for industrialised countries to use cheap carbon offsets to achieve their emissions reduction pledges, and as a consequence, disincentivise their domestic economic transitions. The Agreement did however produce an enhanced adaptation package for developing countries, including a $100 billion per year Green Climate Fund, and technology transfer mechanism.

Largely based on this enhanced adaptation support, negotiations at the Seventeen Conference of the Parties (COP17) held in Durban, South Africa, produced a landmark agreement that saw developed and developing countries (the US and China, critically) agree to establish a new fully inclusive climate agreement by 2015 covering the period between 2020 and 2030 – in other words, a universal successor to the Kyoto Protocol.[154] Negotiations on the text would proceed under the Durban Platform for Enhanced Action with the expectation that the final draft would be delivered at the Twenty-First Conference of the Parties (COP21) to be held in Paris in 2015.

The Eighteenth Conference of the Parties (COP18) was held in Doha, Qatar, in 2012. The COP18 negotiations produced the Doha Amendment in which the Parties would finally agree to commit to emission reductions under the Kyoto Protocol's second commitment period (2013–2020). The Parties also agreed to a strengthened target of 18 per cent below 1990 levels for the second period (up from 5 per cent below 1990 during the first period, 2008–2012).[155] In late 2013, the Nineteenth Conference of the Parties (COP19) was held in Warsaw, Poland. Here the Parties agreed to prepare Intended Nationally Determined Contributions (INDCs) – which were emission reduction pledges expressed as targets out to 2025 or 2030 – and lodge them to the UNFCCC before COP21. INDCs had their origins in the Copenhagen COP where developed countries agreed to provide national mitigation pledges for the period up to 2020 (and developing countries were invited to specify their voluntary 'nationally appropriate mitigation actions'). In effect, the Warsaw COP formalised an extension to the country pledges.

In a major development on the road to Paris, in November 2014, US President Barack Obama and President Xi Jinping of China announced that they had struck a landmark bilateral agreement to reduce carbon emissions. Both Obama and Xi agreed that climate change was 'one of the greatest threats facing humanity' and, on this basis, announced strengthened individual emissions reduction targets, where the US would aim to reduce economy-wide emissions by 26 per cent to 28 per cent below 2005 levels in 2025, while China for the first time offered an absolute emissions reduction target, aiming to have emissions peak by around 2030. Both leaders also formally agreed to work with other countries to forge a 'legal instrument or an agreed outcome with legal force under the Convention applicable to all Parties at the United Nations Climate Conference in Paris in 2015'.[156]

Buoyed by the US-China announcement of post-2020 targets, the Twentieth Conference of the Parties (COP20) held soon after in Lima, Peru, made a specific call for countries to submit their 2025 or 2030 emission reduction pledges – INDCs – to the UN Secretariat by early 2015.[157] The Durban Platform for Enhanced Action process had prepared a 12-page negotiating text comprising all the elements of a post-2020 agreement, and circulated it to the delegates in Lima. By the end of the negotiations however the text had blown out to 37 pages, reflecting the diversity of interests involved.[158] The text would need to be shortened to a more manageable size to improve the chances of a successful agreement in Paris in December 2015.

In addition to the political momentum provided by the US-China agreement, several other background dynamics between 2010 and 2015 supported these outcomes and assisted to build momentum towards COP21 in Paris, including the improved business case for renewable energy expansion (and the opposite for coal), the inescapably worrying temperature and carbon dioxide rises (see Chapter 1), and the increasing pressure for the UN Security Council to consider climate change as a threat to international peace and security (see Chapter 3). Also in this time, the US, UK, and EU and organisations operating under their auspices such as NATO began to publish an increasing number of major reports that sought to link and integrate climate change into security, defence, and foreign policy discussions. For example, Western countries and their affiliates produced 27 climate security reports between 2003 and 2009, compared to 42 between 2010 and 2015 (see Chapter 3).[159]

Adding to the climate security discourse in the lead up to the Paris negotiations, in March 2014, the IPCC published its Fifth Assessment Report (AR5) on change impacts, vulnerability, and adaption. The AR5 contained many more references to conflict compared to earlier IPCC reports, distributed across several chapters including sections on 'observed events', 'emergent risks', and 'Africa'.[160] In addition, for the first time in IPCC history, the AR5 devoted a full chapter to the concept of Human Security, reflecting the major expansion in climate-conflict research since the AR4 of 2007.[161] However, the AR5 delivered mixed messages on the closeness of the relationship. On the one hand, it explained that 'many of the factors that increase the risk of civil war and other armed conflicts are sensitive to climate change' including poverty, economic shocks, weak economic growth, and unstable political institutions; yet on the other hand, 'collectively the research does not conclude that there is a strong positive relationship between warming and armed conflict'.[162] The latter statement reflected the IPCC's growing scepticism about the closeness of the climate-conflict relationship, while the former provided support for climate security advocates.

On 30 November 2015, the hugely anticipated COP21 began in Paris. By this stage, 181 parties had communicated their INDCs to the UNFCCC (with seven more announced during the COP), which meant that a positive outcome was achievable simply by virtue of the delegates accepting the pledges.[163] However, the negotiations sought to produce a far more ambitious result: a viable successor to the Kyoto Protocol that would avoid dangerous climate change. The Secretary-General opened proceedings urging 'bold climate action' in the form of a 'universal, meaningful agreement' to avoid the negative economic, environmental and security consequences associated with 'even a 2-degree rise'.[164]

On 12 December 2015, after almost two weeks of intense negotiations, the Paris Agreement was formally adopted by 196 Parties. The prevailing view at the time was that Paris achieved what Copenhagen could not – an inclusive, binding, and ambitious international treaty to reduce global emissions. The Agreement satisfied the inclusivity criteria by becoming the first international climate agreement that brought together all nations, developed and developing, under the one framework and required them to articulate how they intended to reduce emissions (by contrast, under Kyoto, developed countries had emissions targets while developing countries did not). Article 26 of the Agreement however allowed countries to formally withdraw from the Agreement three years after it had entered into force – which occurred on 4 November 2016 after 55 countries representing 55 per cent of global emissions had ratified it.[165] The Agreement was binding in the sense that it was an international treaty, legally binding on the parties that ratified it.[166] However, the US and emerging economies insisted that the INDCs – comprising country-specific emissions targets – were not legally binding under the Agreement (by contrast, under the Kyoto Protocol country specific targets were binding).[167] The Agreement satisfied the ambitious criteria in so far as it presented a much bolder temperature goal (1.5°C) compared to any previous international climate agreement (by contrast, the Kyoto Protocol did not articulate a temperature limit). The relevant provision, Article 2 (1), asserted that,

> This Agreement, in enhancing the implementation of the Convention, including its objective [to prevent dangerous climate change], aims to strengthen the global response to the threat of climate change … including by: Holding the increase in the global average temperature to well below 2°C above pre-industrial levels and pursuing efforts to limit the temperature increase to 1.5°C above pre-industrial levels, recognizing that this would significantly reduce the risks and impacts of climate change;[168]

However, the temperature goals were aspirational. Indeed, the Agreement did not include any specific targets or dates, or the means for coordinating national emission reduction contributions, to ensure a collective outcome that avoided breaching the 2°C 'dangerous' threshold, let alone 1.5°C. An additional concern was that the collective emissions reductions under the INDC pledges lodged in Paris, even if fully achieved, would still lead to warming of around 2.7°C by 2100.[169] The parties recognised this situation by emphasising 'with serious concern the urgent need to address the significant gap' between existing mitigation pledges (2.7°C) and the Paris temperature goals (1.5°C–2°C).[170] To close the gap, Article 4 of the Agreement stipulated that all parties must submit a new Nationally Determined Contribution (NDCs) every five years with each successive submission expected to 'represent a progression' in mitigation effort.[171]

The R2P's institutional embeddedness in the UN system, and beyond, continued between 2015 and 2022. For example, in 2015 alone, the Security Council passed 12 resolutions that referenced the R2P in country situations (Syria, South Sudan, Sudan, Liberia, Central African Republic, Democratic Republic of Congo).[172] In 2016, 11 more were passed (encompassing new countries, Mali and Somalia). In 2017, the Council passed 12 resolutions referencing R2P (dropping Liberia, but including the Lake Chad Basin, and Great Lakes region). And between 2018 and 2022, the Council passed a further 22 country-related resolutions on R2P. In total, the Security Council passed only two country-specific resolutions referencing the R2P in the period 2001–2010, compared to 74 in the period 2011–2022.[173] Similar patterns of activity can be observed in Security Council Presidential Statements, Human Rights Council resolutions, and General Assembly resolutions.[174] On 1 January 2017, António Guterres replaced Ban Ki-moon as UN Secretary-General and, consistent with his predecessors (Ki-moon and Annan), continued to support R2P's growth and development in the UN system.[175] Beyond the UN, by 2020, more than 60 states and intergovernmental organisations had established R2P 'Focal Points' with high-level officials tasked with analysing atrocity risks and mobilising responses.[176] In addition, R2P strategies had informed government and nongovernment action in preventing and reacting to violence in Kenya, Sierra Leone, Liberia, Guinea, Côte d'Ivoire, The Gambia, Kyrgyzstan, Congo, South Sudan, and the Central African Republic.[177] However, as Alex Bellamy wrote in 2020,

> for all the political and procedural progress made towards its implementation, the governments that willingly *pledged* themselves to R2P have failed where it matters most: in the prevention of atrocity crimes and protection of vulnerable populations (emphasis added). [178]

Failed government pledges to adequately prevent and respond to imminent or occurring atrocity crimes (as the R2P principles call for) contributed to mass atrocities being committed in the post-2011 environment in Afghanistan, Ethiopia, Iraq, Myanmar, Nigeria, South Sudan, Sudan, and Yemen, among other countries, including where R2P-based prevention activities were evident such as in the Congo.[179] Repeated failures to adequately address the government of Syria's use of chemical weapons against its own population is case in point. Leading up to the 2011 uprising and subsequent conflict, the international community had a deep awareness of Syria's chemical weapons assets. In 2012 and early 2013, several small-scale attacks on civilian populations were reported. No action was taken. On 21 August 2013, Syrian government forces launched a major sarin gas attack against civilians in the Ghouta district of Damascus, killing 1,400 people, many of them children.[180]

The international community's reaction was limited to UN Security Council Resolution 2118, unanimously adopted on 27 September 2013, resolving to eliminate Syria's chemical weapons stockpiles. In April 2014, only seven months after the Resolution had passed, Syrian forces launched a series of chemical attacks on the cities of Hama and Idlib and have continued to use such weapons (sarin, chlorine, and mustard gas) in subsequent years. No one has ever been prosecuted for the Ghouta atrocity, which some scholars argue appears emblematic of a flourishing broader culture of impunity that enables atrocities.[181]

The post-Paris climate agenda focused on developing processes and supporting policies that would help close the gap between pledged warming (as reflected in the collective government NDCs) and agreed warming (as reflected in the UN Paris Agreement). The expectation at COP21 in December 2015 was that the Parties would need to submit their first updated NDC – indicating a strengthened emissions reduction target – to the UNFCCC by 2025. However, in January 2016 a 'decision' was reached ('decision' documents function as instructions on how to implement the Agreement but are not legally binding) that requested that the Parties submit a new or updated NDC by 2020.[182] Further signs that the world was focused on emissions reduction included President Obama and President Xi agreeing in September 2016 that their respective countries would ratify the Paris Agreement; and on 4 November, the Agreement itself entering into force having satisfied the threshold criteria (above).[183] However, this momentum on global emissions reductions would soon face a series of significant obstacles.

On 7 November 2016, the Twenty-Second Conference of the Parties (COP22) commenced in Marrakech, Morocco. Two days into negotiations over the Paris 'rule book' – the guidelines necessary to implement the Paris Agreement – Donald Trump was elected US President. Two key outcomes of COP22 reflected a concern about the prospect of stalled action on emissions reductions: the *Marrakech Action Proclamation* and *Marrakech Partnership for Global Climate Action*. Both documents emphasised that the world was 'united', and that the momentum was 'irreversible' on emissions reductions, but also that sub-national actors needed to step up (e.g. state and provincial governments, cities, regions, businesses, and investors).[184] In June 2017, President Trump announced that the US would withdraw from the Paris Agreement.[185]

The key outcome of the Twenty-Third Conference of the Parties (COP23) held in Fiji in 2017 was the launch of the 'facilitative dialogue' (also known as the Talanoa Dialogue). This process, due to commence in 2018, was to facilitate communication between the Parties in the hope that open diplomacy and enhanced transparency might lead to more ambitious emissions targets in the 2020 NDCs. The following year, at the 2018 Conference of the Parties (COP24) held in Katowice, Poland, the Parties adopted much of the Paris 'rule book' (also known as the Katowice Climate Package).[186] The rule book offered the Parties critical guidance on planning, implementing, and reviewing NDCs over each five-year period.[187] However, the final Katowice text allowed countries to use 'nationally appropriate methodologies' for emissions reporting (compared to the common method used under the Paris Agreement) opening up significant loopholes for creative and even dishonest emissions reporting, especially exercised through the land-use sector.[188]

In 2019, despite these and other various UN processes put in place, global enthusiasm for elevated ambition on emission reductions continued to wane. To stave off a growing sense of despondency among climate action supporters (especially among school children who had been leading protests worldwide) and to rebuild momentum for enhanced action under the Paris Agreement, in September 2019, the UN Secretary-General, António Guterres, convened a Climate Action Summit in New York. Despite the emergence of an 'Ambition alliance' of countries committed to raising their national commitments in the NDCs, the Summit ended with none of the high-emitting countries committing to increase their ambition prior to the end of 2020.[189] On 4 November 2019, the Trump Administration gave formal notice of the US's intention to withdraw from the Paris Agreement in 12 months. In addition, Presidents Bolsanaro in Brazil and Putin in Russia and Prime Ministers Modi in India, Morrison in Australia, and Johnson in the UK were all either casting doubt on the scientific basis of climate change or actively working to block efforts to boost ambition and accelerate actions to implement the Paris Agreement.[190]

In December 2019, the Twenty-Fifth Conference of the Parties (COP25) was eventually held in Madrid, Spain, after Brazil – the original confirmed host – withdrew its offer. In the negotiations, countries such as Australia, Saudi Arabia, China, India, and Brazil opposed placing any obligation on countries to submit enhanced emission reduction pledges by 2020, arguing it should be each country's own decision. In the concluding days of the negotiations, a draft decision was circulated that removed any call for countries to 'update' or 'enhance' their climate plans by 2020; instead, only inviting them to 'communicate' them in 2020 – which put no obligation on enhanced ambition.[191] However, stronger language was eventually reinserted that urged the countries to consider the gap between existing pledges and what was needed to limit warming 'well below 2°C'.[192] By the end of 2020, only 44 Parties representing around 28 per cent of global emissions lodged new or updated NDCs to the UNFCCC, the UK a stand out pledging to reduce economy-wide emissions by at least 68 per cent by 2030, compared to 1990 levels.[193] Notable absences were the US, China, and India.[194] On 4 November 2020, the US formally withdrew from the Paris Agreement.[195]

After a 12-month postponement due to the Covid-19 pandemic, in November 2021, the Twenty-Sixth Conference of the Parties (COP26) took place in Glasgow, Scotland. It was the largest COP in history with more than 38,000 registered badge holders. In the lead up to the COP, some Western governments were arguing for a 'green recovery' from the pandemic. Leading the charge was the new US President, Joe Biden, who had already reinstated the US to the Paris Agreement, unveiled a $2 trillion climate-related spending package, and announced a new NDC, pledging to reduce economy-wide emissions by between 50 per cent and 55 per cent below 2005 levels in 2030.[196] In the UK, the Johnson government, having been convinced by the climate science,[197] had announced $21 billion (£12 billion) to drive a 'green industrial revolution'.[198] And the EU, a traditional global climate leader, had announced it would reduce emissions by at least 55 per cent below 1990 levels by 2030 to deliver on the 'European Green Deal'.[199] Receiving less praise, China's NDC reaffirmed its 2020 position that it would aim for emissions to peak before 2030 and achieve carbon neutrality before 2060; and India pledged to achieve net zero by 2070.

The COP26 produced the *Glasgow Climate Pact* (GCP).[200] While not legally binding, the GCP delivered three significant improvements compared to the Paris Agreement. First, it confirmed 1.5°C as the primary global temperature ceiling, shifting emphasis away from 'well below 2°C' – the outer limit set in the Paris Agreement.[201] Second, the GCP plainly stated that limiting global warming to 1.5°C will require 'reducing global carbon dioxide emissions by 45 per cent by 2030 relative to the 2010 level and to net zero around mid-century' – the Paris Agreement makes no mention of 2030 as a target.[202] Third, and the most significant 'first', was the GCP's inclusion of language on constraining unabated coal power and inefficient fossil fuel subsidies.[203] Despite the production and burning of coal, oil, and gas contributing nearly two thirds of global GHG emissions, the Paris Agreement failed to mention the words 'fossil fuels', 'coal', 'oil', or 'gas'. The GCP by contrast, '*Calls upon* Parties to accelerate … efforts towards the phasedown of unabated coal power and phase-out of inefficient fossil fuel subsidies'.[204] The initial wording on coal was 'phase-out', but with pressure from China and India the final text was amended to 'phasedown'. In terms of coal, leading climate scholar Joanna Depledge (and colleagues) explain that 'the GCP represents a genuine breakthrough: this is the first time that a COP decision has referred to constraining coal – or any fossil fuel – indicating a real progression in global opinion'.[205] By the end of COP26, 151 countries had registered new or updated NDCs with the UNFCCC.[206] The UN Environment Programme immediately calculated that these pledges, as they stood, would result in 2.5°C of warming by 2100.[207] The UNFCCC subsequently requested that the Parties 'revisit and strengthen' their NDC emission reduction pledges 'as necessary to align them with the Paris Agreement's temperature goal' and resubmit them by the end of 2022.[208]

In April 2022, the IPCC released its Sixth Assessment Report (AR6) on impacts, vulnerability, and adaptation.[209] The AR6 was again cautious about making a direct link between climate change and conflict stating that 'there is no (scholarly) consensus on the causal association between observed climate change and conflicts',[210] but confirmed that climate change does contribute to increased conflict along indirect pathways such as food and water insecurity and migration, and via intermediate background political and socio-economic factors.[211] The AR6 also identified a variety of climate adaptation policies involving governance, livelihoods, and women that could help prevent conflict and build a sustainable peace.[212] For the Secretary-General, António Guterres, the AR6 revealed 'a litany of broken climate promises' by governments, adding that 'it is a file of shame, cataloguing the empty pledges that put us firmly on track toward an unliveable world'.[213] As of December 2022, only five countries had submitted a stronger NDC pledge to the UNFCCC (Australia, Norway, Singapore, UAE, and Thailand).[214] The UN Environment Programme's latest emissions gap report finds that 'no credible pathway' to achieve 1.5°C is currently in place.[215]

Conclusion

This chapter traced the major international efforts to curb global warming and prevent atrocities. Overall, it showed that the aim of both policy development processes was to protect human populations from harm. However, significant convergence and divergence marked the overall period. The chapter showed that in the period 1989–2001, the global warming mitigation and conflict/atrocity prevention policy tracks, while separate, acknowledged one another in a significant way. For example, the climate policy process saw violent conflict as the result of failed climate action, while the atrocity prevention policy development process saw unmitigated global warming and environmental degradation as a contributing factor to violent conflict and human misery. However, in the period 2001–2011, the conflict/atrocity prevention policy development process seemed to shear off climate impacts from its discussion. This could be due to a combination of factors including most of the R2P leadership group and the UN Membership, especially those from the Global South, seeking a narrow definition on the R2P scope, the climate-sceptic Bush Administration taking over The White House (2001–2008), and the IPCC's increased scepticism about the relationship between climate and conflict. In the period 2011–2022, the climate action and R2P agendas became increasingly embedded in their respective UN processes, while at the same time, governments worldwide for the most part neglected their pledges to reduce emissions and prevent atrocities. However, Western governments especially (US, UK, and EU), individually and collectively via the UN Security Council, began to increasingly integrate climate action policies into foreign, defence, and security strategies. The next chapter begins by detailing this recent policy history, before examining some of the conceptual parallels between the climate action agenda and the Responsibility to Protect.

Notes

1 David M. Crowe, *War Crimes, Genocide and Justice: A Global History* (New York: Palgrave Macmillan, 2014), p 164 and 192.
2 Adam Jones, *Genocide: A Comprehensive Introduction*, 3rd Edition (London: Routledge, 2017), p 11–12. The Genocide Convention has significantly influenced international and domestic laws since, for example, the definition of

genocide has been adopted by the International Criminal Court and has been incorporated into domestic laws in several countries.

3 'The Geneva Conventions of 1949 and their Additional Protocols', *International Committee of the Red Cross* (accessed 2022).

4 Rachel Carson, *Silent Spring* (London: Houghton and Mifflin, 1962).

5 Peter Christoff and Robyn Eckersley, *Globalization and the Environment* (Maryland: Rowman & Littlefield, 2013), p 35.

6 Ibid.

7 'Declaration of the United Nations Conference on the Human Environment', Point 3, 16 June 1972.

8 'Report of the International Conference on the Assessment of the Role of Carbon Dioxide and of Other Greenhouse Gases in Climate Variations and Associated Impacts', Villach, Austria, 9–15 October 1985, p 10.

9 *Our Common Future*, Report of the World Commission on Environment and Development (Oxford: Oxford University Press, 1987). See also, Jim MacNeill, 'The Greening of International Relations', *International Journal*, Vol 45, No 1, 1989/1990, p 3–4.

10 Christoff and Eckersley, *Globalization and the Environment*, p 56.

11 'Peace, Security, Development, and the Environment', in *Our Common Future,* 1987, Point 2.

12 Ibid., Point 6 and 7.

13 *The Changing Atmosphere: Implications for Global Security*. Toronto, Canada, June 27–30, WMO, Geneva, 1988.

14 Ibid., see 'attendees' in Appendix.

15 Ibid., p 16.

16 Ibid., p 292.

17 Ibid., p 296.

18 IPCC, 1990: *Climate Change: The IPCC Impacts Assessment*. Report Prepared for IPCC by Working Group II (Canberra: Australian Government Publishing, 1990). See Chapter 1 of the present book for IPCC reports on the Physical Scientific Basis – Working Group 1.

19 Ibid., section 5 p 4.

20 IPCC, 1992: *Climate Change 1992: The Supplementary Report to the IPCC Impacts Assessment* (Canberra: Australian Government Publishing, 1993), see, for example, p 6, 33–34, 51, and 77.

21 'History of the Convention', *UNFCCC* (accessed 2021).

22 'UNFCCC: 25 Years of Effort and Achievement', *UNFCCC* (accessed 2022).

23 'United Nations Convention on Climate Change' (FCCC/INFORMAL/84 GE.05-62220 (E) 200705), *UNFCCC*, 1992, p 4.

24 'What is the United Nations Framework Convention on Climate Change?', *UNFCCC* (accessed 2022).

25 'Two Degrees: The History of Climate Change's Speed Limit', *Carbon Brief*, 8 December 2014.

26 Ibid.

27 Chapter VIII. Consideration of questions under the responsibility of the Security Council for the maintenance of international peace and security. Decision of 31 January 1992 (3046th meeting): Statement by the President, p. 821.

28 *The Responsibility of The Security Council in The Maintenance of International Peace and Security*, Decision of 31 January 1992 (3046th meeting): Statement by the President.

29 Boutros Boutros-Ghali, *An Agenda for Peace: Preventive Diplomacy, Peacemaking and Peacekeeping* (New York: United Nations, 1992). Note: Gareth Evans describes this report at 'pathbreaking' in the development of the R2P.
30 Ibid., p 6–7.
31 Ibid.
32 Ibid., p 8.
33 Ibid.
34 Resolution 827 (S/RES/827). Adopted by the Security Council at its 3217th meeting on 25 May 1993.
35 Crowe, *War Crimes, Genocide and Justice*, p 342 and 349.
36 Ibid., p 343.
37 Ibid.
38 UNDP, *Human Development Report 1994: New Dimensions of Human Security* (Oxford University Press, 1994).
39 Ibid.
40 *The Responsibility to Protect*, Report of the International Commission on Intervention and State Sovereignty (Ottawa: International Development Research Centre, 2001), p 13.
41 UNDP, *Human Development Report 1994*, p 22–38.
42 'Kyoto Protocol to the United Nations Framework Convention on Climate Change', *United Nations*, 1998, p 3.
43 'What is the Kyoto Protocol?', *UNFCCC* (accessed 2022).
44 'Two Degrees', *Carbon Brief*, 8 December 2014.
45 IPCC, 1995: *Climate Change 1995: Impacts, Adaptations and Mitigations of Climate Change*. Contribution of Working Group II to the Second Assessment Report of the Intergovernmental Panel on Climate Change (Cambridge and New York: Cambridge University Press, 1996).
46 Ibid.
47 IPCC, Special Report 1998: *The Regional Impacts of Climate Change: An Assessment of Vulnerability. Summary for Policymakers*, November 1997, p 3–4.
48 Francis M. Deng, Sadikiel Kimaro, Terrence Lyons, Donald Rothchild, and I. William Zartman, *Sovereignty as Responsibility: Conflict Management in Africa* (Washington, DC: Brookings Institution Press, 1996).
49 Francis Deng in communication with Gareth Evans – see Gareth Evans, *The Responsibility to Protect: Ending Mass Atrocity Crimes Once and For All* (Washington, DC: Brookings Institution Press, 2008), p 273.
50 Roberta Cohen had previously written on this issue in 1991 that 'sovereignty carries with it a responsibility on the part of the governments to protect its citizens' in *Human Rights Protection for Internally Displaced Persons*, Refugee Policy Group, Washington, 1991.
51 Roberta Cohen and Francis Deng, 'Exodus within Borders: The Uprooted that Never Left Home', *Foreign Affairs*, Vol 77, No 4, 1998, p 12–16.
52 Francis M. Deng et al., *Sovereignty as Responsibility*, p 161.
53 Ibid., p 121, 164, and 174.
54 Crowe, *War Crimes, Genocide and Justice*, p 349–350.
55 Tony Blair, 'Doctrine of the International Community', Speech, Chicago Economic Club, 22 April 1999.
56 Ibid.
57 Ibid.

58 Kofi Annan, 'Two Concepts of Sovereignty', *The Economist*, 16 September 1999.
59 Ibid.
60 Kofi Annan, '*We the Peoples': The Role of the United Nations in the 21st Century*, UN Millennium Report (United Nations, 2000).
61 Ibid.
62 Ibid., p 40–46.
63 Ibid., p 34.
64 *The Responsibility to Protect*, ICISS.
65 Evans, *The Responsibility to Protect*, p 41–42.
66 'Basic Principles', in *The Responsibility to Protect*, ICISS, p xi. A different formulation is expressed – that adds a responsibility for the international community to assist the state in question – in Alex J. Bellamy, *The Responsibility to Protect: A Defense* (New York: Oxford University Press, 2015) p 2–3.
67 *The Responsibility to Protect*, ICISS.
68 Ibid., p 33.
69 IPCC 2001: *Climate Change 2001: Impacts, Adaptation, and Vulnerability*. Contribution of Working Group II to the Third Assessment Report of the Intergovernmental Panel on Climate Change (Cambridge: Cambridge University Press, 2001), p 397.
70 'UNFCCC: 25 Years of Effort and Achievement', (accessed 2022).
71 Michael Grubb and Farhana Yamin, 'Climate Collapse at the Hague: What Happened, Why and Where do we go from Here?', *International Affairs*, Vol 77, No 2, 2001, p 261–276.
72 See 'Foreword' by the UN Secretary-General Kofi Annan in *A More Secure World: Our Shared Responsibility*. Report of the Secretary-General's High-level Panel on Threats, Challenges and Change (United Nations, 2004), p vii.
73 Ibid.
74 Ibid., p 2.
75 Ibid., p 26.
76 Ibid., p 44–45.
77 Ibid., p 30.
78 Crowe, *War Crimes, Genocide and Justice*, p 359–360.
79 Jones, *Genocide: A Comprehensive Introduction*, p 714–715.
80 'Cambodia: UN-backed Tribunal Ends with Conviction Upheld for Last Living Khmer Rouge Leader', *UN News*, 22 September 2022.
81 'Khmer Rouge: What did a 16-year Genocide Trial Achieve?', *BBC*, 22 September 2022.
82 Kofi Annan, *In Larger Freedom: Towards Development, Security and Human Rights for All*. Report of the Secretary General, A/59/205 (United Nations, 2005).
83 Evans, *The Responsibility to Protect* (Brookings, 2008), p 45.
84 Annan, *In Larger Freedom*, 2005, p 59.
85 Ibid., p 20.
86 Ibid., p 21.
87 Evans, *The Responsibility to Protect* (Brookings, 2008), p 47.
88 *2005 World Summit Outcome*, General Assembly Resolution, A/RES/60/1, 24 October 2005.
89 Evans, *The Responsibility to Protect*, p 47. See also, Bellamy, *The Responsibility to Protect*, p 5.

90 Edward C. Luck, '*Environmental Emergencies and the Responsibility to Protect: A Bridge Too Far?*', Proceedings of the Annual Meeting (American Society of International Law), Vol 103, 2009, p 33.
91 Ibid., p 33.
92 *2005 World Summit Outcome*, p 11–15.
93 Resolution 1664, S/RES/1674, Security Council, 28 April 2006. See also Bellamy, *The Responsibility to Protect*, p 7.
94 'UNFCCC: 25 Years of Effort and Achievement' (accessed 2022).
95 Nils Petter Gleditsch and Ragnhild Nordas, 'Conflicting Messages? The IPCC on Conflict and Human Security', *Political Geography*, Vol 43, 2014, p 84.
96 Security Council, 5663rd Meeting, S/PV.5663, New York, 17 April 2007. See also 'Climate Security at the UNSC – A Short History', *Climate Diplomacy* and *Adelphi* (accessed 2022).
97 Ibid., p 2.
98 CNA Military Advisory Board, *National Security and the Threat of Climate Change* (Alexandria, VA: CNA Corporation, 2007).
99 Ibid., p 3.
100 Gleditsch and Nordas, 'Conflicting Messages?', p 84.
101 Ibid.
102 'COP 13: Bali Climate Change Conference 2007: Full Text: Ban-Ki Moon Speech', *The Guardian*, 12 December 2007.
103 Peter Christoff, 'The Bali Roadmap: Climate change, COP 13 and Beyond', *Environmental Politics*, Vol 17, No 3, June 2008, p 468.
104 'Bali Action Plan', Decision 1/CP.13, FCCC/CP/2007/6/Add.1, *UNFCCC*, 14 March 2008, p 3. See also, 'Two Degrees', *Carbon Brief*, 8 December 2014.
105 Ibid.
106 Christoff, 'The Bali Roadmap', p 467–468.
107 Bellamy, *The Responsibility to Protect*, p 8.
108 Resolution 1706, S/RES/1706, Security Council, 31 August 2006.
109 Bellamy, *The Responsibility to Protect*, p 8.
110 Evans, *The Responsibility to Protect* (Brookings, 2008), p 51.
111 Bellamy, *The Responsibility to Protect*, p 8.
112 Gareth Evans, 'The Responsibility to Protect: Meeting the Challenges', Lecture, Singapore, 5 August 2008. See also, Evans, *The Responsibility to Protect* (Brookings, 2008), p 65–66.
113 Bellamy, *The Responsibility to Protect*, p 13.
114 Ibid., p 8.
115 Evans, 'The Responsibility to Protect: Meeting the Challenges'. See also, Evans, *The Responsibility to Protect* (Brookings, 2008), p 51.
116 Luck, 'Environmental Emergencies and the Responsibility to Protect', p 33.
117 Ban Ki-moon, *Implementing the Responsibility to Protect*, Report of the Secretary-General, A/63/677, General Assembly, 12 January 2009.
118 Ibid., p 1.
119 Ibid., p 8.
120 Resolution 63/308, *The Responsibility to Protect*, A/RES/63/308, General Assembly, 14 September 2009.
121 Resolution 1894, S/RES/1894 (2009), Security Council, 11 November 2009.
122 Proposed by the Pacific Island states Federated States of Micronesia, Fiji, Kiribati, Nauru, Palau, Papua New Guinea, Republic of the Marshall Islands, Samoa, Solomon Islands, Tuvalu, Tonga, and Vanuatu.

123 Resolution 63/281, A/RES/63/281, General Assembly, 3 June 2009, p 2.
124 Ban Ki-moon, *Climate Change and its Possible Security Implications*, Report of the Secretary-General, A/64/350, 11 September 2009, p 2.
125 *The Responsibility to Protect*, ICISS, p 23. See also, Peter Wittig, Permanent Representative Germany, 'Concept Note', Letter dated 1 July 2011 from the Permanent Representative of Germany to the United Nations addressed to the Secretary-General, UN Security Council, 5 July 2011, p 2.
126 Ban Ki-moon, *Climate Change and its Possible Security Implications*, A/64/350, p 6.
127 *The Responsibility to Protect*, ICISS, p 23.
128 Peter Christoff, 'Cold Climate in Copenhagen: China and the United States at COP15', *Environmental Politics*, Vol 19, Issue 4, 2010, p 637–656.
129 'Copenhagen Accord', Decision 4/CP.15, FCCC/CP/2009/11/Add.1, *UNFCCC*, 30 March 2010, p 5.
130 Christoff, 'Cold Climate in Copenhagen, p 651.
131 Kali Robinson and Will Merrow, 'The Arab Spring at 10 Years: What's the Legacy of the Uprisings?', *Council on Foreign Relations*, 3 December 2020.
132 Bellamy, *The Responsibility to Protect*, p 9.
133 The International Criminal Court estimated that between 500 and 700 civilians were killed in February 2011 – see 'First Report of The Prosecutor of The International Criminal Court to the UN Security Council Pursuant to UNSCR 1970', *International Criminal Court*, 2011, p 4.
134 Simon Adams, 'Libya', in Alex J. Bellamy and Tim Dunne (Eds.), *The Oxford Handbook of the Responsibility to Protect* (New York: Oxford University Press, 2016), p 770.
135 'Ban Strongly Condemns Qadhafi's Actions Against Protesters, Calls for Punishment', *UN News*, 23 February 2011.
136 Resolution 1970, S/RES/1970, Security Council, 26 February 2011. See also, Adams, 'Libya', in Bellamy and Dunne (Eds.), *The Oxford Handbook of the Responsibility to Protect*, p 769.
137 Resolution 1973, S/RES/1973, Security Council, 17 March 2011.
138 Bellamy, *The Responsibility to Protect*, p 10.
139 Gareth Evans, 'Learning from Libya', *Project Syndicate*, 8 April 2016.
140 See Justin Morris, 'Libya and Syria: R2P and the Spectre of the Swinging Pendulum', *International Affairs*, Vol 89, No 5, September 2013, p 1265.
141 Ban Ki-moon, 'Remarks at Breakfast Roundtable with Foreign Ministers on "The Responsibility to Protect: Responding to Imminent Threats of Mass Atrocities"', 23 September 2011.
142 Gareth Evans, 'End of the Argument: How We Won the Debate Over Stopping Genocide', *Foreign Policy*, 28 November 2011.
143 Adams, 'Libya', in Bellamy and Dunne (Eds.), *The Oxford Handbook of the Responsibility to Protect*, p 772.
144 Ibid., p 773.
145 Kofi Annan, presentation in Ottawa reported in 'Don't Fear Islamic Resurgence from Arab Spring, Former UN Chief Tells Students', *The Globe and The Mail* (Canada), 4 November 2011.
146 Evans, 'Learning from Libya'.
147 'Libya letter by Obama, Cameron and Sarkozy: Full text', *BBC*, 15 April 2011.
148 Adams, 'Libya', in Bellamy and Dunne (Eds.), *The Oxford Handbook of the Responsibility to Protect*, p 774.

149 UNSC resolutions in this period: Libya × 5, Central African Republic × 4, South Sudan × 3, Syria × 2, and Cote d'Ivoire, Yemen, Mali, Somalia × 1 – see 'R2P References in United Nations Security Council Resolutions and Presidential Statements', *Global Centre for R2P* (accessed 2022).
150 Evans, 'Learning from Libya'.
151 'Syria War: Average of One Child Injured or Killed Every Eight Hours Over Past 10 Years', *UNICEF*, 12 March 2021.
152 'Secretary-General's remarks to Climate Change Conference (UNFCCC COP16 High-Level Segment)', Statements, UN Secretary-General, Cancun, 07 December 2010.
153 'The Cancun Agreements: Outcome of the Work of the Ad Hoc Working Group on Long-term Cooperative Action Under the Convention', Decision1/CP.16, FCCC/CP/2010/7/Add.1, *UNFCCC*, 15 March 2011.
154 'Essential Background – Durban Outcomes', *UNFCCC* (accessed 2022).
155 'The Doha Amendment', *UNFCCC* (accessed 2022).
156 'U.S.-China Joint Announcement on Climate Change', Press Release, Office of the White House (from Beijing), 12 November 2014. See also 'US and China Strike Deal on Carbon Cuts in Push for Global Climate Change Pact', *The Guardian*, 14 November 2014.
157 Robyn Eckersley, 'Endgame at Lima', *The Interpreter* (Lowy Institute), 13 December 2014.
158 Ibid.
159 'Climate Security 101', *The Centre for Climate and Security* (accessed 2022).
160 Gleditsch and Nordas, 'Conflicting Messages?', p 84.
161 IPCC, 2014: *Climate Change 2014: Impacts, Adaptation, and Vulnerability. Part A: Global and Sectoral Aspects*. Contribution of Working Group II to the Fifth Assessment Report of the Intergovernmental Panel on Climate Change (Cambridge and New York: Cambridge University Press, 2014), p 758–791. See also, Gleditsch and Nordas, 'Conflicting Messages?', p 84.
162 Gleditsch and Nordas, 'Conflicting Messages?',, p 85.
163 Peter Christoff, 'The Promissory Note: COP 21 and the Paris Climate Agreement', *Environmental Politics*, Vol 25, No 5, 2016, p 765–787.
164 Ban Ki-moon, 'Speech to COP21 Leaders Summit', *Secretary-General*, 30 November 2015.
165 Don Henry and Ben L. Parr, 'Trump's Political Climate', *Pursuit*, 5 June 2017.
166 Christoff, 'The Promissory Note', p 775.
167 Ibid.
168 'The Paris Agreement to the United Nations Framework Convention on Climate Change', *United Nations*, 12 December 2015, p 3.
169 A. Admiraal et al. 2015 in Peter Christoff, 'The Promissory Note', p 775.
170 'Adoption of the Paris Agreement: Proposal by the President', Draft Decision –/CP.21, FCCC/CP/2015/L.9/Rev.1, *UNFCCC*, 12 December 2015, p 2.
171 'The Paris Agreement', 12 December 2015, p 4–5.
172 'R2P References in United Nations Security Council Resolutions and Presidential Statements', *Global Centre for R2P* (accessed 2022).
173 Ibid.
174 See, for example, 'UN General Assembly Resolutions Referencing R2P', *Global Centre for the R2P* (accessed 2022).

175 Charles T. Hunt, Cecilia Jacob, and Adrian Gallagher, 'Progress, Problems, and Prospects: R2P 15 Years After the World Summit', *Global Responsibility to Protect*, Vol 12, 2020, p 360.
176 Gareth Evans, 'R2P: The Dream and the Reality', Lecture, European Centre for the Responsibility to Protect, Leeds University, 26 November 2020. See also, Simon Adams, 'Responsibility to Protect and the World in 2020', *Global Centre for the R2P*, 10 February 2020; 'Members of the Global Network of R2P Focal Points', *Global Centre for R2P* (accessed 2020).
177 Evans, 'R2P: The Dream and the Reality'.
178 Alex J. Bellamy, 'The Responsibility to Protect at 15: A Promise Unfulfilled', *Global Centre for the R2P*, 21 September 2020.
179 Adams, 'Responsibility to Protect and the World in 2020'.
180 'Syria: Eighth Anniversary of the Ghouta Chemical Weapons Attack', Press Statement, *US State Department*, 21 August 2021.
181 See Bellamy, 'The Responsibility to Protect at 15'.
182 'Decisions Adopted by the Conference of the Parties: Adoption of the Paris Agreement', Decision 1/CP.21, FCCC/CP/2015/10/Add.1, *UNFCCC*, 29 January 2016, paragraph 24 (NDC) and paragraph 20 (facilitative dialogue).
183 'Breakthrough as US and China Agree to Ratify Paris Climate Deal', *The Guardian*, 4 September 2016.
184 'Marrakech Partnership for Global Climate Action', *UNFCCC*, November 2016; 'Marrakech Action Proclamation Expresses Irreversible Momentum on Climate', *UNFCCC*, November 2016.
185 'Trump Will Withdraw U.S. From Paris Climate Agreement', *The New York Times*, 1 June 2017.
186 'Navigating the Paris Rulebook', *World Resources Institute* (accessed 2022).
187 Ibid.
188 'COP24: Key Outcomes Agreed at the UN Climate Talks in Katowice', *Carbon Brief*, 16 December 2018.
189 Peter Newell and Olivia Taylor, 'Fiddling While the Planet Burns? COP25 in Perspective', *Globalizations*, Vol 17, No 4, 2020, p 581.
190 Ibid.
191 Ibid.
192 Ibid.
193 'United Kingdom of Great Britain and Northern Ireland's Nationally Determined Contribution', UK Government, 2020, p 1. See also, 'UK Vows to Outdo Other Economies with 68% Emissions Cuts by 2030', *The Guardian*, 4 December 2020.
194 Josh Gabbatiss, 'Analysis: Which Countries Met the UN's 2020 Deadline to Raise "Climate Ambition"?', *Carbon Brief*, 8 January 2021.
195 'Climate Change: US Formally Withdraws from Paris Agreement', *BBC*, 4 November 2020.
196 'The United States of America. Nationally Determined Contribution. Reducing Greenhouse Gases in the United States: A 2030 Emissions Target', US Government, 2021, p 1. See also, 'Biden Announces $2 Trillion Climate Plan', *The New York Times*, 1 February 2021; 'Biden Returns US to Paris Climate Accord Hours After Becoming President', *The Guardian*, 21 January 2021.
197 'Climate Change: The Science Briefing that Convinced Boris Johnson', *BBC*, 2 February 2022.

198 'Policy Paper: The Ten Point Plan for a Green Industrial Revolution', UK Government, 18 November 2020.
199 'Delivering the European Green Deal', European Commission (accessed 2022).
200 'Decisions Adopted by the Conference of the Parties Serving as the Meeting of the Parties to the Paris Agreement', Decision 1/CMA.3, Glasgow Climate Pact, FCCC/PA/CMA/2021/10/Add.1, *UNFCCC*, 8 March 2022. See also, 'What Were the Outcomes of COP26?', UK Parliament, 27 January 2022; 'The Glasgow Climate Pact – Key Outcomes from COP26', UNFCCC (accessed 2022).
201 Joanna Depledge, Miguel Saldivia, and Cristina Peñasco, 'Glass Half Full or Glass Half Empty?: The 2021 Glasgow Climate Conference', *Climate Policy*, Vol 22, No 2, 2022, p 147–157.
202 Ibid., p 148.
203 Ibid., p 149–151.
204 'Glasgow Climate Pact', FCCC/PA/CMA/2021/10/Add.1, paragraph 36.
205 Depledge et al., 'Glass Half Full or Glass Half Empty?', p 150.
206 Helen Mountford, David Waskow, Lorena Gonzalez, Chirag Gajjar, Nathan Cogswell, Mima Holt, Taryn Fransen, Molly Bergen, and Rhys Gerholdt, 'COP26: Key Outcomes from the UN Climate Talks in Glasgow', *World Resources Institute*, 17 November 2021.
207 Ibid.
208 Ibid.
209 IPCC, 2022: *Climate Change 2022: Impacts, Adaptation and Vulnerability*. Contribution of Working Group II to the Sixth Assessment Report of the Intergovernmental Panel on Climate Change (Cambridge and New York: Cambridge University Press, 2022).
210 Ibid., p 593.
211 Mary Potts, Lukas Rüttinger, and Janani Vivekananda, 'What Does the IPCC Report Tell us About Climate and Conflict?', *Climate Diplomacy*, 7 March 2022.
212 Ibid.
213 António Guterres, 'Secretary-General Warns of Climate Emergency, Calling Intergovernmental Panel's Report 'a File of Shame', While Saying Leaders 'Are Lying', Fuelling Flames', Statements and Messages, *United Nations*, 4 April 2022.
214 'CAT Climate Target Update Tracker', *Climate Action Tracker* (accessed 2022).
215 'Climate Change: No 'Credible Pathway' to 1.5C limit, UNEP Warns', Meeting Coverage and Press Releases, *United Nations*, 27 October 2022.

3 Climate Change and R2P

Towards Integration

Chapter 2 examined the broader history and relationship between climate action policy processes and conflict/atrocity prevention processes (1988–2022). It showed that while periods of convergence and divergence marked the timeframe, a general trend towards closer relations can be observed overall.

The first section of this chapter builds on this discussion by examining in closer detail recent Western efforts (2010–2022), individually and in collaboration through the UN Security Council, to integrate climate change impacts and policies into conflict/atrocity prevention and response policies. Ultimately, the section aims to provide a springboard to begin to examine the relationship between climate change and the Responsibility to Protect (R2P). Chapter 2 showed that throughout the 2000s, integration attempts were sparse, but increasing, and mostly concerned with unpacking the relationship between climate impacts and peace and security at a global and national level. In the period leading up to the Paris Agreement (2010–2015), discussed below, Western countries hastened integration attempts and began to shift their attention towards the relationship between climate impacts and specific national defence strategies, and to a lesser extent, foreign policies. In the aftermath of the Paris Agreement (2015–2020), Western governments redoubled their integration efforts, and in another shift, increasingly began to explicitly address the relationship between climate policies (not just climate impacts) and broader conflict prevention and resolution policies (not just defence and foreign policies). With the incoming Biden Administration in the US (December 2020-ongoing), integration efforts accelerated again with actionable climate-conflict policies identified and cautious links made between climate impacts and atrocity prevention frameworks. Overall, the section shows that as of 2022, integration is at a high point in the historical record.

The second section begins to directly address the relationship between climate change and the R2P. It does so by introducing some of the key conceptual debates in the R2P literature that relate to climate change. First, it examines the prominent debate over the R2P's four-fold scope. Woven into the discussion are three ancillary conceptual issues (namely, the challenge of proving

DOI: 10.4324/9780367201364-5

intent, North–South relations, and human security) as well as four proposals to integrate climate change into the R2P scope (namely, include climate change in the existing R2P crimes; eliminate R2P's use of force option; establish ecocide as a crime under the Rome Statute; and/or launch a new international commission on the R2P). Next, this section details and critiques the R2P's five-fold scope for assessing whether a country should be considered one of 'R2P concern'. This discussion includes insights into the threat-multiplier thesis, the relationship between countries vulnerable to atrocities and climate impacts, and climate projections. It suggests that 'climate vulnerability' is deserving of its own analytical space and representation in the applied (and scholarly) R2P literature.

Integrating Climate with Peace and Security – The Recent Background

Throughout the 2000s, UN and Western efforts to integrate climate change issues into peace and security issues slowly gathered pace. For example, in 2007, the UN Security Council held its first formal discussion on the interlinkages between climate and security, which was conducted under the UK's Council Presidency (see Chapter 2).[1] Independently of the UN, in the mid-2000s, the US, UK, and EU defence, intelligence, and foreign agencies released more than a dozen reports framing climate change as a 'threat multiplier' with national and international security implications; and on several occasions, emphasised the need to improve prevention and preparedness capabilities at home and abroad to address climate-related disasters and conflict.[2] However, the reports tended to focus on unpacking the multiple risks posed by climate change, rather than on policy solutions or policy integration.

Between 2010 and 2015 (the year the Paris Agreement was reached) Western countries continued to generate analysis and reports integrating climate (impacts, primarily) and security (defence primarily, but also foreign policies). The UK, as part of the EU in the 2010s, not only contributed to EU policy development (discussed below), but also produced its own national climate security strategies that drew on expertise from across the UK defence and security landscape (e.g. the UK Strategic Defence and Security Review of 2010, Climate Change Risk Assessment of 2012, Global Strategic Trends out to 2045 of 2014). In the US, national security agencies released multiple reports each year on the relationship between climate and security.[3] But overall, in this period, similarly to the 2000s, the UK and the US largely focused on improving their understanding of the national security risks associated with a variety of climate impacts.[4]

The EU provided the most innovative approaches to integration in this period. Notably, in 2011, the EU launched its landmark 'climate diplomacy' initiative that began to engage in a more comprehensive fashion on the foreign policy dimensions of climate change.[5] At the time, the notion of EU 'climate diplomacy' covered three broad themes: supporting UNFCCC negotiations, promoting low-emission development, and reducing climate security risks – which, the EU argued, would require the involvement of EU foreign, defence and external economic policies, as well as development cooperation, humanitarian aid, and all regional and bilateral relations (in 2013, the EU Foreign Affairs Council described this process as 'mainstreaming' climate security into all external policies).[6] However, consistent with the previous decade, and the US and UK reports, much greater emphasis was placed on understanding the threat compared to developing integrated policy solutions.

Integration also became increasing central to UN Security Council discussions in this period – vigorously led by Western countries. In early July 2011, Peter Wittig, Germany's Permanent Representative to the UN and President of the UN Security Council, called for an Open Debate on the impacts of climate change on the maintenance of international peace and security.[7] Wittig's 'concept note' to inform the debate described the threat of climate change as 'already tangible' and only going to get worse 'in the years to come', which meant, he argued, that 'the effects of climate change thus go beyond the mandate' of the UNFCCC. Wittig's note also identified a range of policy responses that suggested a need to integrate climate change impacts (e.g. resource scarcity) into traditional human protection policies operationalised through the UN such as 'conflict analysis, mission planning and mission monitoring' and 'peacebuilding activities'.[8]

On 20 July 2011, the debate was held in the UN Security Council. UN Secretary-General Ban Ki-moon opened proceedings by arguing, similarly to Wittig, that climate change 'not only exacerbates threats to international peace and security, it *is* a threat to international peace and security', and given this, the involvement of the UN Security Council in the climate debate is 'not only appropriate – it is essential'.[9] However despite these overtures, the Statement by the President of the Security Council, the key outcome document, downplayed the threat that climate change posed, describing it as a possible future concern that may 'aggravate certain existing threats to international peace and security'; and based on this, the Security Council reaffirms that the UNFCCC should be the primary instrument to address climate change.[10] However, the Statement indicated a receptiveness to climate change analysis overlaying traditional conflict analysis should it enhance Council decision-making.[11]

This debate was followed by two Security Council 'Arria Formula' meetings (informal gatherings to discuss issues of importance, but do not constitute a formal security council activity), held in 2013 and mid-2015 on the security implications of rising temperatures and climate change. The 2013 discussions resulted in Pacific Island states whose 'very existence' was threatened urging the Security Council to consider climate change a threat to international peace and security.[12] The 2015 meeting, co-hosted by the Permanent Missions of Spain and Malaysia to the UN, explored the interconnectedness between climate change and international peace and security. The outcome statement explained that national governments are increasingly seeking to integrate climate change into their security and foreign policy options – a trend that should be reflected at an international security level.[13] The Arria Formula discussions confirmed that among the permanent members, the UK, the US, and France, generally viewed the Council as an appropriate vehicle to address climate change as a part of its conflict prevention efforts, while Russia and China did not.[14]

In the aftermath of the Paris Agreement, between 2015 and 2020, Western governments – independently and collaborating through the UN – redoubled their efforts on integration, and increasing began to explicitly address the relationship between climate policies (not just climate impacts) and broader conflict prevention and resolution policies (not just defence and foreign policies). The EU would once again lead the discussion. For example, in 2016, the EU's Global Strategy on Foreign and Security Policy would repeatedly describe climate change as a threat multiplier and a root cause of conflict, but as was becoming chronic, offered little by way of actionable policy responses, rather making several broad brush remarks about the need for a coherent and holistic response, including with regard to conflict prevention activities in climate exposed states.[15] However, the 2016 report did represent a step forward by circling around the conflict prevention-climate action nexus.

The EU's 2017 report on integrated external action, which covered climate security, began to drill down into policy suggestions to not only prevent climate-related violent conflict but also rebuild in its aftermath through enhanced diplomatic engagement, policy dialogue and analysis that 'systematically included climate and environmental indicators in conflict early warning systems'.[16] However, climate adaptation strategies that would have relevance, such as supporting climate-exposed sectors (e.g. agriculture, fisheries and critical infrastructure), were addressed separately to this discussion. Another major step-change would come in 2018. In February, the European Council resolved to 'further mainstream the nexus between climate change and security in policy dialogue, conflict prevention, development and humanitarian action and disaster risk strategies'.[17] In June, this promise materialised with the EU Parliament adopting a comprehensive resolution on 'climate diplomacy' that included an expanded explanatory paragraph related to climate-informed conflict prevention and resolution:[18]

> (The Resolution) Highlights that EU foreign policy should develop capacities to monitor climate change related risks, including crisis prevention and conflict sensitivity; believes that consequential and rapid climate action contributes essentially to the prevention of social, economic, but also security risks, the prevention of conflicts and instabilities and ultimately the prevention of major political, social and economic costs; stresses, therefore, the importance of mainstreaming climate diplomacy in the EU conflict prevention policies, broadening and adapting the scope of EU missions and programmes in third countries and conflict areas.[19]

The major limitation of the Resolution – as with previous EU outputs – involved translating the high-level commitments and mainstreaming requirements into specific and actionable climate-informed policies on peace, humanitarian issues and development.[20]

In 2017, the UN Security Council held a third Arria Formula debate on climate security, themed: 'preparing for the security implications of rising temperatures'.[21] The discussion sought to uncover the opportunities to better institutionalise climate security across the UN system, for example, by requiring the UN Security Council to integrate climate impacts into its traditional conflict assessment and prevention measures – with the suggestion that even the Responsibility to Protect agenda should provide inspiration on how to achieve UN reform in the interests of human protection (see *The Responsibility to Prepare*).[22] Also in 2017, the Council passed Resolution 2349 on the Lake Chad Basin, which in a first, featured language on R2P and climate change. On the former, the Resolution '*Reiterates* the primary responsibility of Member States to protect civilian populations on their territories, in accordance with their obligations under international law'. On the latter, the Resolution

> *Recognises* the adverse effects of climate change and ecological changes among other factors on the stability of the Region, including through water scarcity, drought, desertification, land degradation, and food insecurity, and *emphasises* the need for adequate risk assessments and risk management strategies by governments and the United Nations relating to these factors.[23]

Similar R2P-climate formulations would appear in six other Security Council resolutions between 2018 and 2021.[24] And climate change would feature independently of R2P in 16 other Security Council resolutions between 2017 and 2021.[25]

In 2018, two new UN-based initiatives were created that sought to address climate-related security risks. First, the Group of Friends on Climate and Security (GFCS), which comprised more than 50 UN member states (including the US from 2021) and the UN's regional groups. The GFCS was tasked with raising the political profile of climate security risks and developing integrated policy solutions. Second, the Climate Security Mechanism (CSM) was established, which was located in and staffed primarily by the UN Department of Political and Peacebuilding Affairs (DPPA) but drew on expertise from the UN Development Programme and the UN Environment Programme. The CSM was tasked with developing tools that integrated climate change into the DPPA's existing conflict prevention and response capabilities such as early warnings frameworks and conflict mediation.

Also in 2018, the Intergovernmental Panel on Climate Change (IPCC) released a Special Report on the impacts of 1.5°C of global warming compared to 2°C.[26] While acknowledging that some research had found a limited relationship between climate change and deadly conflict, compared to previous IPCC reports, the Special Report conveyed a strengthened message about the relationship, reporting that climate change impacts have been present in most human conflicts across the world; and as the temperature climbs (from

1.5°C to 2°C and beyond) 'the rates of human conflict could increase'.[27] This report informed the climate security agenda at the UN Security Council in the following years. For example, in 2019 the Council again debated the impacts of climate-related disasters on international peace and security, with an unprecedented number of Member States taking the floor, many at a ministerial level. The Under-Secretary-General of the UN DPPA, Rosemary DiCarlo, opened proceedings, stating that,

> A report released by the Intergovernmental Panel on Climate Change last October predicted more heat waves, heavier rain events, higher sea levels and more severe damage to agriculture. These trends represent a security risk for the entire world. However, their consequences are felt most strongly in regions that are already vulnerable, where climate change and extreme weather compound existing grievances and threats.[28]

A core part of the solution, the DPPA chief continued, was to develop 'an integrated risk assessment framework to analyse climate-related security risks' in collaboration with practitioners from across the UN and beyond.[29]

The momentum for integrating climate change into conflict prevention tools extended into 2020, with the German Foreign Minister and Security Council President, Heiko Maas, announcing at the conclusion of yet another Security Council debate on this topic, the creation of an Informal Expert Group (IEG) on climate-related risks to peace and security.[30] In November 2020, the IEG met for the first time with all 15 Council members in attendance to discuss operational responses to climate-related security risk, especially those in Somalia.[31] The upshot of the meeting was the appointment of an Environmental Security Advisor for the UN mission to Somalia – a first for the UN.[32]

The EU's next major policy development also came in November 2020 with the EU's Climate Change and Defence Roadmap.[33] The Roadmap aimed to better integrate climate factors into the EU's defence policies (dissimilarly from 'climate diplomacy' that also encompassed foreign, economic, and development policy domains). The roadmap was successful in the sense that it presented an expanded list of actionable climate security policy responses under its 'operation dimension' – for example, through the development of a climate-conflict early warning system and mainstreaming climate into its 'crisis management toolbox' for civilian and military missions.[34] However, the roadmap suffered from few specific priorities, actions and timelines for reform – and not much by way of tangible outcomes, which led some to argue that the roadmap was just another example of the EU remaining 'more concerned with institutional mandates, capacities, and agenda setting' and failing to 'move on to a more action orientated stage'.[35]

Between 2021 and 2022, the Western-led integration agenda (now led by the incoming Biden Administration) took major strides, which included cautiously identifying climate impacts in new atrocity prevention frameworks. In December 2020, the US announced that it too would 'mainstream' climate change across government agencies, with appointees to key posts (e.g. National Intelligence, State Department, Defense, Treasury, Homeland Security, UN representative) conveying strong concerns about the threat posed by climate change and the need for a strong policy response.[36] Soon after, in January 2021, the President signed an Executive Order to 'put the climate crisis at the centre of US foreign policy and national security'.[37] In August, the US special presidential envoy for climate change (a new position established under the EO), John Kerry, argued that responding to climate change is 'without exaggeration, about survival' requiring 'a response that is comparable to wartime mobilization'.[38]

On 21 October 2021, four major reports on climate change were published by core US national security and foreign policy components: The Department of Homeland Security, National Intelligence Council, Defense Department, and The White House.[39] The homeland security report was naturally inward looking, expressing a strong intention to deliver improved inter-agency capability to prevent and respond to domestic natural disasters, for example. The intelligence and defence reports were similar in the sense that they both focused on the variety of threats posed by climate change across the world – now and into the future – especially their geostrategic implications. However, they stopped short of offering detailed policy responses, rather simply suggesting that the physical impacts of climate change globally will likely add additional demands on US diplomatic, economic, humanitarian and military resources, therefore closer integration across US policymaking domains should be pursued.

The most impressive output was the rather innocuously titled *Report on the Impact of Climate Change on Migration*, produced by The White House itself.[40] The report was divided into three substantive sections. First, it presented a sophisticated understanding of the relationship between climate change and violent conflict with a focus on providing 'individual-level human security'. Second, it presented a variety of policies, operationalised across the departments of State, Defense, Agriculture and USAid, to prevent as a priority, but also respond to, climate-forced displacement by establishing food insecurity early warning systems, investing in Disaster Risk Reduction programmes with a focus on nature-based solutions, developing training and education facilities to diversify livelihoods, strengthening healthcare systems, supporting human rights protections and building inclusive governance capacities with an active civil society in climate hotspots. Third, it explored in detail the legal strategies, conventions, agreements, and instruments that the US could pursue to protect and resettle displaced persons, both in cross-border migration situations and those internally displaced. The argument went that while climate-forced migration alone will not necessarily trigger a US response, when combined with violence or armed conflict – which is likely, the report suggests – the US will have a duty to protect vulnerable populations and resettle them in a post-conflict environment. Therefore, overall, the report was especially novel because it framed the complex, and in some instances increasing distant, relationship between climate change and violent conflict as an advantage rather than disadvantage – that is, the intervening political, social, and economic variables, regularly seen by climate determinists as a block on policy action, as offering the US multiple entry points to respond to climate-related human protection situations.

In a major break from the past, in September 2021, the EU published the action-orientated *Concept for an Integrated Approach on Climate Change and Security.*[41] The Concept was the EU's opening gambit towards establishing a 'framework for operational approaches' to address external security risks by mainstreaming climate change aspects across relevant policy fields, with an emphasis on crisis and conflict management. The 'multi-phased' proto framework identified policy options across 'conflict prevention', 'crisis response', and 'conflict resolution and stabilisation'. Conflict prevention was the Concept's top priority, with the main policy to develop a fully integrated climate-enhanced conflict early warning system with processes in place to translate early warning into an effective response. Crisis response (also labelled 'peacebuilding' and described as actioning 'stabilisation efforts in situations where conflict has already erupted') involved mainstreaming climate aspects into the EU's regular defence and security planning and operations, and more tangibly, supporting a 'host State's' (suggesting consent is required) climate-related law enforcement efforts and security related to environmental governance. Conflict resolution and stabilisation involved equipping conflict mediators with climate-related knowledge (and expanding EU capacity in this regard), including climate risks and opportunities in existing Disarmament, Demobilisation and Reintegration processes, and incorporating climate-related security risks in Security Sector Reform processes. Other areas touched upon included integrating climate change into relevant humanitarian aid, human rights, education and training programmes. As with The White House report, providing human security was the focus.

The UK delivered new but limited integrated initiatives in 2021 (having formally ceased to be a member of the EU since January 2020). For example, in March 2021, the UK Ministry of Defence published a roadmap on how to address the impacts of climate change on the activities of the UK's armed forces.[42] The report represented an important change in attitude towards the acceptance of climate security as core to military operations.[43] However, the focus was on examining climate-related impacts rather than identifying policy opportunities, and indeed was inward looking rather than seeking a whole-of-government approach.[44] Also in March 2021, the UK released its landmark integrated review of security, defence, development and foreign policy titled, *Global Britain in a Competitive Age*.[45] The review bundled together previous separate areas of external policy and set out the UK's objectives through to 2030, which included, as one of the four priority areas, to help build 'global resilience' by tackling climate change and development.[46] However, on this, the review presented seemingly inconsistent messages by strongly supporting multilateral emissions reduction forums such as the Paris Agreement (indirect human protection) while also continuing with cuts to foreign aid for climate-vulnerable regions such as the Middle East and North Africa (direct human protection).[47] In addition, despite conveying concern about the security implications of climate change on several occasions, no integrated policy solutions were offered. However, the review did represent a significant step towards the UK adopting a framework akin to the EU's 'climate diplomacy' (above).

Overlaying this uptick of domestic activity in 2021, Western governments also continued to press ahead on the need for climate security integration through the UN Security Council. In February 2021, the UK as Council President called for a high-level debate on the topic.[48] In a letter to inform discussions addressed to UN Secretary-General António Guterres, the UK Permanent Representative to the UN, Barbara Woodward, wrote that climate change is expected to increasingly exacerbate conflict risk factors and therefore, 'conflict prevention and peacebuilding approaches will increasingly need to integrate climate considerations into their strategies'.[49] The debate was attended by all 15 Council members, including many heads of state and government. As Chair of the debate, UK Prime Minister, Boris Johnson, declared that 'it is absolutely clear that climate change is a threat to our collective security and the security of our nations', and therefore it is 'the duty of this UN Security Council to act'.[50] President of France, Emmanuel Macron, and US Special Envoy on Climate, John Kerry, delivered similar sentiments.

In September 2021, the Security Council under Ireland's Presidency, held another debate on the security implications of climate change. During the meeting several Council Members requested a thematic resolution on climate and security, which was subsequently drafted. In December 2021, a thematic resolution – co-sponsored by 113 member states – on systematically integrating climate-related security risks into the UN's conflict prevention, conflict management and peacebuilding work was put to a vote in the Security Council.[51] In the lead-up to the vote, Council members Estonia, France, Ireland, Kenya, Mexico, Niger, Norway, Tunisia, Saint Vincent and the Grenadines, the UK, the US, and Vietnam had expressed support for a more systematic integration of climate-related security risks into the Council's work. By contrast, China, India and Russia objected to the draft resolution, arguing that there is 'no clear scientific background for equating climate change with security concerns'.[52] The draft resolution vote was 12 in favour (the above 12) and 2 against (Russia, India) with 1 abstention (China), which meant that the Council rejected the draft owing to a negative vote by a permanent member (Russia). This division between the P5 – the US, the UK, and France (and EU countries broadly) versus Russia and China (and India) – over the Council's role in addressing climate change was present in all previous climate security debates.[53]

In 2022, the Biden Administration continued to integrate climate change into US national security planning and policies, for example, in January through the National Defense Authorization Act, in February with the announcement of the US Army's first Climate Strategy, and in October under the US National Security Strategy.[54] However, the April 2022 release of a 'prologue' to the 2020 Global Fragility Strategy is most representative of the US's growing intent to provide an overarching framework to prevent and respond to climate-related violent conflict.[55] By way of a brief backstory, in 2019, the US Congress passed the Global Fragility Act directing the State Department (under the Trump Administration at the time) to deliver a new integrated framework to guide US conflict prevention and resolution activities in fragile contexts abroad. In 2020, the *United States Strategy to Prevent Conflict and Promote Stability* was released.[56] The Strategy – drawing on the capacities present in the departments of state, defence, development, and treasury, among others – identified a variety of tools and policy initiatives to prevent violent conflict as a priority, covering diplomatic and foreign assistance (e.g. promote dialogue and mediation, protect human rights, tackle corruption, and governance reforms), security (e.g. support public order and build foreign security force capacity), economics and trade (e.g. support private sector investment and employment, and consider targeted and broad-based sanctions) and legal options (e.g. integrating the White House-led Atrocity Early Warning Task Force, established under the Elie Wiesel Global and Atrocity Prevention Act, into the Strategy). The trouble was that 'climate change' was not mentioned once.[57] The April 'prologue' addressed this deficiency by discussing the contribution climate change makes to fragility and conflict; and suggested the need to 'test' climate-conflict policy responses.[58]

In addition to these wide-ranging reports, in 2022, the US and UK both published reports that focused on new integrative approaches to atrocity prevention and resolution – the *United States Strategy to Anticipate, Prevent, and Respond to Atrocities* of July 2022; and the UK's *Preventing Future Mass Atrocities Around the World* of October 2022.[59] Both reports built on previous atrocity prevention frameworks in their respective countries (e.g. the Obama Administration's 2016 Executive Order 13729; and the 2019 UK Approach to Preventing Mass Atrocities),[60] and reaffirmed the centrality of atrocity prevention and response in government policy. The omnipresent task identified in both reports was to develop a 'whole-of-government', 'integrated', 'inter-agency (US)', 'cross-departmental' (UK) policy approach to address atrocities. Both reports explained that this process had recently begun under new initiatives to coordinate government action (the White House-led Atrocity Prevention Task Force; and the UK Office for Conflict, Stabilisation and Mediation – especially its 'conflict and atrocity prevention hub'). The reports were intended to be initial steps towards establishing an integrated policy toolkit; however, both presented a substantial list of strategies and policy options under consideration (covered in recent US and UK reports above). Both reports cautiously identified climate change as a possible driver of atrocities. The US atrocity prevention strategy states, very cautiously, that it 'aligns with the prevention architecture articulated in the US Strategy to Prevent Conflict and Promote Stability' (above), which means that climate change may be included in a future US atrocity framework. The UK report, comparatively less cautiously, explicitly identified climate change as a driver of mass atrocities and suggested that there was a need to integrate climate considerations into the development of

UK atrocity prevention and response policies.[61] The EU's most recent atrocity prevention and response framework, titled the *EU Responsibility to Protect – Atrocity Prevention Toolkit*, published in 2018, presents a similar level of integration by identifying 'climate change' as one of 17 'structural risk factors' (and indicators for analysis) that may increase the risk of mass atrocity situations.[62]

This policy development narrative provides a springboard to begin to delve into the R2P–climate action relationship. The section to follow, which rounds out Part I, addresses some of the relevant conceptual debates found in the R2P literature, before Part II of the book examines the operational parallels.

Conceptual Considerations

R2P's Scope and Climate-Related Emergencies

The debate over the scope and limits of the R2P can be traced back to the release of the 2001 *International Commission on Intervention and State Sovereignty* (ICISS) report. In the Commission's view, international military intervention, via Security Council authorisation, was justified to protect human populations should the following 'conscience-shocking' situations occur or appear imminent: genocide, ethnic cleansing, crimes against humanity, other war-related crimes, or 'overwhelming natural or environmental catastrophes, where the state concerned is either unwilling or unable to cope, or call for assistance, and significant loss of life is occurring or threatened'.[63] However, in the period between the release of the 2001 ICISS report and the 2005 World Summit to commemorate the 60th anniversary of the UN, the notion that an R2P-inspired military intervention could be mounted in response to a climate-related human protection catastrophe was dropped from the R2P scope. In the two short paragraphs on the R2P in the World Summit Outcome Document, the narrower four-fold scope was expressed in full on four occasions, including in the title, *The Responsibility to Protect populations from genocide, war crimes, ethnic cleansing, and crimes against humanity*[64] – leaving the General Assembly, who would vote on the Outcome Document at the Summit, with no ambiguity about what situations would trigger an R2P response.

As Chapter 2 suggested, the change of heart between 2001 and 2005 could have been the result of multiple intervening variables, not least the centrality of US Ambassador to the UN, John Bolton (representing a climate-sceptic Bush Administration), to the World Summit negotiations. However, the abandonment of climate-related catastrophes was more likely the result of pressure coming from the broader UN Members, as Edward Luck, who served as the UN Secretary-General's Special Adviser on the Responsibility to Protect between 2008 and 2012, recalled in a 2008 speech on environmental emergencies and R2P:

> He (Jean Pin – UNGA President in September 2005) was there up to the eve of the September 2005 agreement on the Outcome Document that included

> RtoP, so he had an unusually close and sustained view of the negotiations. If you actually go back and look at the comments from member states in the spring of 2005 about RtoP, many are decidedly negative. Nevertheless, a few months later they were able to agree on rather detailed consensus language on RtoP. Why? … According to Jean Pin, the critical point in the negotiations came when Munir Akram, then the Pakistani Permanent Representative (to the UN), not usually described as a great fan of RtoP, insisted that the delegates needed to know exactly which cases and situations they were talking. The ICISS report had really been rather broad on this score, not defining the scope of RtoP crimes in a specific or consistent manner. Member states, on the other hand, wanted to delineate the scope of RtoP's application as precisely as possible. So they were able to negotiate these four crimes ... In many discussions with member states (in the years that followed), it became very clear that what they were really talking about in 2005 was the question of mass violence. Broader matters of governmental neglect and malfeasance were well beyond the scope of the 2005 agreement.[65]

Luck suggests therefore that one of the key reasons that the R2P proposal made it through to the World Summit Outcome Document, which was not assured, was because it was eventually confirmed, and accepted by the Member States, that the R2P's scope *would not* include climate-related, or any other, human protection issues beyond the four atrocity crimes. The World Summit's approval of the R2P's four-fold scope however would not mark the end of the debate – in fact, quite the opposite.

Those who continue to support the R2P's four-fold scope do so primarily on the basis of two interrelated arguments: first, that an international consensus was reached at the 2005 World Summit, and second, to stretch the R2P scope beyond the four crimes would be to the detriment of its practical utility. For example, as UN Secretary-General Ban Ki-moon argued in 2009:

> The responsibility to protect applies, until Member States decide otherwise, only to the four specified crimes and violations: genocide, war crimes, ethnic cleansing and crimes against humanity. To try to extend it to cover other calamities, such as HIV/AIDS, climate change or the response to natural disasters, would undermine the 2005 consensus and stretch the concept beyond recognition or operational utility.[66]

Similarly, Gareth Evans argues that a key misconception about the application of the R2P is that it 'covers all human protection issues', including natural disasters, and by seeking to apply the R2P too broadly 'in non-atrocity contexts, is to dilute to the point of uselessness its role as a mobiliser of instinctive, universal action in cases of conscience-shocking killing, ethnic cleansing and other such crimes against humanity'.[67] R2P scholar, Alex Bellamy, agrees that widening the R2P scope 'would likely undermine the international consensus on which it is built and would make it more difficult to find clarity in relation to implementation'.[68]

Proponents of an expanded version of the R2P's scope – one that includes climate-related human protection catastrophes – deploy similar arguments. But instead of pointing to the international consensus marshalled at the 2005 World Summit, they point to the 12 months international consultation period, which included eleven regional roundtables and national consultations across five continents, among many other high-level discussions and meetings, that produced the R2P's foundational document, the 2001 ICISS report.[69] Former Canadian Foreign Minister, Lloyd Axworthy, who initiated the ICISS process, has since written extensively on the need to be true to the original 2001 report and expand the R2P scope to include environmental calamities. For example, in response to the devastating 2008 Cyclone Nargis, which caused the death of 130,000 people and affected 2.4 million more, Axworthy wrote,

> If R2P does not apply to situations where a government is actively working to deprive large numbers of people of life-saving assistance, then we must ask how far the international community has come in saying that it will never again sit idly by in the face of mass human catastrophe. Indeed, the 2001 report of the International Commission on Intervention and State Sovereignty anticipated a situation such as the one in cyclone-ravaged Burma when it included in its threshold criteria 'overwhelming natural or environmental catastrophes, where the state concerned is either unwilling or unable to cope, or call for assistance, and significant loss of life is occurring or threatened'.[70]

The 2008 cyclone in Myanmar acted as a catalyst for debate on the appropriateness of the R2P scope. The French Foreign Minister, Bernard Kouchner, spurred this debate with his call soon after the cyclone hit for the UN Security Council to authorise the delivery of aid without the consent of the government of Myanmar:

> We are seeing at the United Nations whether we can implement the Responsibility to Protect, given that food, boats and relief teams are there, and obtain a United Nations' resolution which authorizes the delivery (of aid) and imposes this on the Burmese government.[71]

For Axworthy, Kouchner's call for an R2P-type response 'clearly demonstrated that threats to people come not just at the point of a gun; they can also be caused by severe negligence in the aftermath of natural disasters or the outbreak of rampant disease'.[72] By contrast, Evans wrote that Kouchner should understand that R2P 'is not about human security generally, or protecting people from the impact of natural disasters ... Rather, [it] is about protecting vulnerable populations from genocide, war crimes, ethnic cleansing and crimes against humanity'.[73] Prominent R2P author, Roberta Cohen, has also been critical of the UN's commitment to a narrow scope, including in response to Cyclone Nargis, writing that the Secretary-General Ki-moon's remarks on R2P ruled out its potential to protect 'the millions of persons expected to be uprooted by disasters and climate change'.[74]

However, despite these differences, both Evans and Axworthy agree that R2P-inspired intervention is feasible if climate-related human catastrophes *also* involve one or more of the four R2P crimes. For example, Evans wrote in response to the 2008 Cyclone that 'the short answer is that natural disasters, as such, are not R2P situations, but they can be if mass atrocity crimes are also involved, and this may have been the case here'.[75] Similarly, Axworthy wrote that while he thought that 'elements of R2P could and should apply in Burma', regardless of that 'the case could have been made that the actions of the military junta were indeed a crime against humanity', triggering a response.[76] This argument however is intertwined with one of the most vexing issues in the R2P literature: proving 'intent'.

Proving that a state's actions or inactions resulted in a mass atrocity situation is notoriously difficult – apart from in the most egregious atrocity situations such as the centrally planned and implemented Rwandan Genocide. According to Ed Luck, a determination of intent 'would be at least as problematic in situations of environmental emergencies' because of the diffuse nature of responsibility for causing those emergencies, for example, Luck continues:

> would you take the government of Bangladesh to task because it has perpetually been unable to stop recurring mass floods after cyclones and typhoons in its neighbourhood? What about desertification? Undoubtedly there has been enormous displacement, suffering, and loss of life in some parts of the world because of desertification. But who is responsible? Surely, not just the governments of the countries where this has occurred.[77]

The Myanmar case provides a good example of this debate in the R2P literature. Joanna Harrington argues that 'Myanmar's restriction of foreign aid in the initial days of the cyclone's aftermath was insufficient to establish that Myanmar's leaders had a requisite intent to commit crimes against humanity, and thus trigger the invocation of the R2P principle'.[78] By contrast, Ashley McLachlan-Bent and John Langmore conclude that 'intentional or not', given interpretations of the Rome Statute, the Myanmar government's 'inability and unwillingness to respond effectively did constitute a crime against humanity, satisfying the criteria needed to invoke the principle of R2P'.[79] Despite the nuanced perspectives, the authors agree that coercive intervention in this situation would have been inappropriate, which ultimately means that they support the status quo on R2P's four-fold scope.

North-South tension, another key conceptual (and practical) issue for the R2P, has found its way into the debate about whether climate change should be included in an expanded R2P scope.[80] As Evans explains, 'A further problem with stretching the R2P concept to embrace what might be described as the whole human security agenda' (including climate change) is that it might be viewed 'as giving an open invitation for the countries of the North to engage to their hearts content in "civilizing missions" that so understandably anger those in the global South, who have experienced it all before'.[81] Evans adds that to dispel fears in the global south that the R2P is merely a neo-imperialist's Trojan horse requires ongoing efforts to clarify that the R2P scope only applies to the four crimes.[82] Mely Caballero-Anthony and Belinda Chng (2009) who write on the 2008 Myanmar case from an Asian perspective (a region where Evans says 'there was, frankly, never much enthusiasm to start with' for the R2P), express support for Evans's assertion about the anxieties that an expanded R2P scope might cause the Asian region especially. However, they offer a different solution. Rather than clarifying the existing scope, they argue for the creation of an 'R2P-Plus' for the Asian region, which, they explain, would be 'responsive to different kinds of human security threats, such as those caused by intentional state neglect in times of natural disasters [and] focuses on the preventive aspects of R2P'.[83] In other words, they suggest that Asian countries would likely support an expanded climate-informed R2P scope were coercive military options dropped from its operational framework.

The human security concept is intertwined with the R2P concept, especially as it relates to the R2P scope. As Chapter 2 showed, the climate action and atrocity prevention agendas converged under the banner of human protection as emphasised by the concept of human security in the *Human Development Report 1994*. But convergence soon gave way to divergence. Subsequently, many climate researchers wholeheartedly adopted human security as the principal lens through which to view climate change and violent conflict – only very recently downscaling it, leaving some to lament that 'human security is somewhat of an idea looking for a home' with few actors interested in its institutionalisation or involvement in the emerging climate security discussion.[84] By contrast, R2P policymakers had several immediate reservations. First, as Evans explains, human security failed 'in answering questions as to who should do precisely what and when in response to an emerging mass atrocity situation'.[85] Similarly, Edward Luck describes human security as 'an admirable way of thinking about security but one that has little operational content'.[86] Second, R2P advocates were concerned that bundling all the major global problems into the R2P's orbit (and scope) – including protecting populations from HIV/AIDs, weapons of mass destruction, and climate change – would render the concept ineffective, or as Evans puts it: 'if R2P is to be about protecting everybody from everything, it will end up protecting nobody from anything'.[87]

It is unclear where R2P's four-fold crimes originate, or why they were selected over other crimes.[88] However, they do almost mirror the four Crimes Against Peace under the 1998 Rome Statute – notwithstanding a mismatch between 'crimes of aggression' and 'ethnic cleansing'.[89] Writing on the links between R2P and ecological intervention, leading climate scholar, Robyn Eckersley, suggests that ecocide events 'may be considered especially "conscience-shocking" because the effects are irreversible', concluding that a moral case may exist for extending R2P's scope to prevent ecocide situations arising.[90] The crime of ecocide – defined as intentional and systematic acts of destruction, degradation and demolishment of ecosystems and specific environments as a consequence of particular types of human activity – was included in UN debates and drafting exercises around the Rome Statute throughout the 1990s, but it was removed from the final text.[91] Increasingly dramatic climate impacts, combined with a better grasp on the links between climate and insecurity, have led to a resurgence of calls for the Statute to be amended to make ecocide the fifth Crime Against Peace.[92] For example, in late 2019, the Maldives and Vanuatu, two nations severely threatened by rising sea levels, urged that an amendment to the Statute be discussed to criminalise acts that amount to ecocide.[93] France, Spain, Finland, and Belgium, along with a variety of high-profile climate activists, conveyed support for such an amendment.[94] Should the Statute be amended, the R2P scope may come under pressure to be amended also.

In an even more radical proposal, Axworthy and Rock argue that instead of amending the R2P scope, it might be time to establish a new international commission to consider how the R2P concept can be mobilised to help manage a range of new threats with 'climate change as an exemplar'.[95] This proposal suggests a complete climate-rendering of the R2P concept is required. The use of military force is not specifically identified as a policy option, but the authors do explain that in 'extreme cases' UN Security Council authorisation may need to be sought to 'preserve or rescue an environmental asset in which there is clearly a shared and essential interest, or prevent environmental degradation that would have significant implications for humanity'.[96] This approach would thus appear to deliver an R2P scope that would allow for military intervention to prevent ecocide or 'crimes against nature' involving no direct harm to humans – which Eckersley identifies as 'the most challenging case because it appeals to a moral referent object beyond humanity'.[97]

The next section investigates the relationship between climate change and R2P's system for identifying countries of R2P concern.

Identifying Countries of R2P Concern

Throughout 2022, leading conflict prevention and resolution INGO, the International Crisis Group (ICG), regularly identified and monitored more than seventy conflict situations globally covering roughly as many countries.[98] But not all these countries involve R2P situations and therefore can be classified as countries of 'R2P concern', because, as Evans explains, R2P situations 'are those where mass atrocity crimes – involving genocide, ethnic cleansing or other war crimes or crimes against humanity – are actually occurring or imminently about to occur', or could foreseeably occur, triggering the need for appropriate preventive measures to be pursued.[99] Substantiating the former situations (occurring or imminent) can be relatively straightforward; however the second, 'could foreseeably occur', presents a far more challenging task.

To help policymakers forecast how a given situation may evolve, Evans identifies 'five factors' in particular that policymakers 'should take into account in making the call that a particular country, in situations where mass atrocities are not obviously occurring right at that time, is indeed one of R2P concern' (and should be placed on an 'R2P Watch List').[100] First, whether the country in question has a history of past atrocities ('past atrocities'). Second, whether the tensions that gave rise to the past conflict or atrocity situation persist ('continuing tensions'). Third, the strength of the country's coping mechanisms to help resolve tensions ('coping mechanisms'). Fourth, the receptivity of the country to external involvement ('external receptiveness'). Fifth, the quality of the country's political leadership ('leadership quality').[101] Evans acknowledges that this is not an exhaustive risk evaluation list, but by applying these factors, he argues, combined with a detailed case-by-case analysis, 'helps narrow down candidates where an atrocity situation could occur' and therefore helps make the case that a particular country is one of R2P concern and that appropriate responses should follow.[102]

A particular country's vulnerability to climate impacts ('climate vulnerability') does not appear as a factor. A possible explanation for this situation could be that most R2P authors view climate change as a 'threat multiplier' – which posits that climate change interacts with and may exacerbate existing threats to peace and security.[103] For example, Evans states that 'there is unquestionably a general causal connection between (climate change and deadly conflict), at least in the sense that climate change is a "threat multiplier"'.[104] Similarly, the Director of the Asia-Pacific Centre for the R2P, Alex Bellamy (and colleagues), explain that 'climate change is less a direct cause of atrocity crimes than a potential trigger or threat multiplier, where there are already existing risk factors for atrocity crimes'.[105] Successive Directors of the Global Centre for the R2P, Simon Adams (2011–2021), and Savita Pawnday (Current) similarly argue that climate change can be a threat multiplier for atrocities in situations where climate impacts, such as water scarcity, interact with existing social, economic, and political challenges, such as ethnic divisions.[106] This is not unusual, indeed the threat multiplier thesis is, to date, the dominant lens globally through which to view and understand the relationship between climate impacts and conflict (and atrocities) – first emerging in the US in 2007,[107] and since readily used by the US, EU, and UK governments and their respective security agencies, as well as the UN, NATO, and the African Union, to name but a few.[108] However, the unintended consequence of using this thesis is that it tends to push climate impacts into the background compared to other drivers and factors associated with conflict – in short, making it appear less significant.[109]

Acknowledgement of such downplaying is reflected in what appears to be a recent shift away from the threat multiplier thesis towards a recognition that climate change should be viewed as a threat to peace and security in its own right. Historical analysis shows that in the 1990s, the climate–conflict relationship in the scholarly literature was understood in simplistic and reductionist terms often under the 'resource wars' banner. From the late 2000s, the more sophisticated notion that climate change exacerbated existing conflict risks and dynamics – 'the threat multiplier thesis' – took precedent.[110] However, some researchers now argue that the threat multiplier thesis is too broad to be useful in practice.[111] While others suggest that it should be viewed only as an entry point to climate security issues because the threat posed by climate change is so pervasive and grave[112] – with some going a step further labelling climate change a 'hyperthreat'.[113] In the applied policy space, it is perhaps telling that the flurry of climate security documents emerging from the US in 2021 (above) do not explicitly use the term 'threat multiplier' at all.[114] More specifically, critical analysis suggests the threat multiplier thesis may not necessarily be the best fit for the R2P concept. First, the thesis channels policy action towards prevention options, becoming somewhat redundant should the conflict enter R2P's reaction and rebuilding phases.[115] Second, it frames security as a state-centric concept, which stands in contrast to the individual human protection focus of the R2P.[116] Third, it strikes a pessimistic tone by focusing on 'risk' and 'threat' minimisation, in comparison to the more optimistic and solutions-orientated R2P framework.[117] In short, including 'climate vulnerability' as an R2P factor may be assisted by adopting a more progressive approach to the climate–conflict relationship.

Perhaps a more compelling justification for elevating 'climate vulnerability' to parity can be demonstrated by examining the relationship between the countries of R2P concern and those vulnerable to climate impacts. On the one hand, the Global Centre for the R2P, a New York-based R2P advocacy group, identifies 36 countries that have experienced or are at risk of experiencing one or more of the four R2P crimes.[118] Twenty-one of the countries are located in North and West Africa (Libya, Sudan and South Sudan, Somalia, Ethiopia, Eritrea, Mali, Burkina Faso, Niger, Cameroon, Nigeria, Guinea, Côte d'Ivoire), sub-Saharan Africa (Central African Republic, Democratic Republic of Congo, Burundi), and the Middle East and West Asia (Yemen, Syria, Iraq, Afghanistan, Pakistan). UN Security Council resolutions that have referenced R2P cover a similar suite of countries and regions (Sudan/South Sudan, Central African Republic, Democratic Republic of Congo, Syria, Libya, Somalia, Mali, Côte d'Ivoire, Yemen, Burundi, Lake Chad Basin and Great Lakes region).[119] Broader conflict analysis undertaken by the International Crisis Group shows a similar geographical distribution.[120]

On the other hand, since 2015, the University of Notre Dame's 'Global Adaptation Index' (ND-GAIN) has ranked more than 180 countries based on their vulnerability to climate impacts.[121] To generate the rank, researchers assess the extent to which climate hazards could negatively impact each country's 'life supporting' systems including food, water, health, ecosystem services, human habitat, and infrastructure. In 2020, the latest ranking, the countries identified as the top ten most vulnerable out of the 182 ranked were, in order (from most to least vulnerable): Niger, Somalia, Chad, Guinea-Bissau, Sudan

(South Sudan), Liberia, Mali, Central African Republic, Eritrea, Rwanda (with the next 20 most vulnerable countries including the Democratic Republic of Congo, Uganda, Sierra Leone, Ethiopia, Yemen, Afghanistan, Burundi, Zimbabwe, Malawi, Burkina Faso, The Gambia).[122] In short, the countries identified as of 'R2P concern' are, by and large, the same as those identified as the most vulnerable to climate impacts.

Congruence between the countries of R2P concern and those vulnerable to climate impacts can also be demonstrated through the distribution of relevant UN field personnel. For example, climate security expert, Florian Krampe, explains that 'ten out of 21 ongoing UN peace operations are located in countries ranked as most exposed to climate change' and 'of the 92,159 personnel deployed to UN peace operations, 74,396 personnel (80 per cent) were deployed in such countries'.[123] Similarly, the UN DPPA explains that

> climate change has major implications for our ability to prevent conflict and sustain peace around the world. This is true in contexts where DPPA manages special political missions, the majority of which are deployed in highly climate vulnerable countries according to the ND-GAIN Index.[124]

The International Committee of the Red Cross is also adapting their operations in part based on 'sixty per cent of the twenty countries considered to be most vulnerable to climate change are sites of armed conflict'.[125]

With this overlap in climate and R2P countries in mind, the most recent climate projections make for especially sobering reading, and an additional justification for inclusion. Global average surface temperatures are expected to exceed 1.5°C by 2035 at the latest, regardless of emissions reductions globally.[126] By the end of this century, warming scenarios range from 1.1°C to 5.7°C above pre-industrial times, dependent of a variety of factors including the pace of the global transition away from fossil fuels, deforestation rates, and feedback loops such as permafrost melt.[127] To put this outer scenario in context, the last global Ice Age about 20,000 years ago saw global average temperatures only 5°C cooler than today.[128] Current country pledges to reduce emissions, if implemented in full, would result in about 2.4–2.8°C of global warming over this century.[129] But a substantial and widening gap between what countries have pledged and global emissions persists. The physical impacts of exceeding temperature thresholds will become more extreme over time. For example, in the future, a 1 in 50-year extreme temperature event could occur roughly: every 6 years at 1.5°C; every 3–4 years at 2°C; and nearly every year at 4°C.[130] The emissions already in the atmosphere, and locked into the global economic system, guarantees that the impacts of climate change will also continue to threaten the security of already vulnerable human populations.[131] For these reasons, among others, world leaders are increasingly conveying extreme concern for the future of humanity.[132] For example, in the last 24 months UN Secretary-General António Guterres has described the current state of climate inaction as 'digging our own graves', 'mutually assured destruction', 'suicidal', 'sleeping walking to climate catastrophe', 'coming to a point of no return' and 'a life-or-death struggle' for survival.[133]

It is a truism that there are many factors and variables to consider when taking stock of a potential atrocity situation, take for example the Darfur atrocity situation in 2003, where scholars and practitioners have given differing weighting to a long list of background and foreground variables, and there interlinkages – for example, colonisation; Cold War rivalry; identity politics; social and economic marginalisation; drought and desertification; and the Government of Sudan's personalities, policies and rhetoric. However, drawing on the latest science, climate–conflict researchers seem increasingly comfortable to describe climate change as 'not simply just another variable'.[134]

Conclusion

This chapter was divided into two main sections. The first section tracked the West's growing efforts to integrate climate-related issues into the traditional peace and security agenda. It showed that in the period 2010–2015 (pre-Paris), integration largely paired climate impacts with defence and security concerns, with foreign policies a junior partner. However, in the period 2015–2020 (post-Paris), climate policies began to increasingly accompany climate impacts, and integration shifted to broader conflict prevention and resolution policies. In the period 2021–ongoing (Biden era), integration accelerated with actionable policies identified and initial steps investigated into the relationship between climate impacts and the atrocity prevention and response agenda. This section aimed to provide an up-to-date account of integration efforts related to the R2P and provide a springboard into the R2P-climate discussion itself.

The second section began to formally explore the relationship between the R2P and climate action agendas. Initially, it addressed a range of conceptual issues related to expanding the scope of the R2P from genocide, crimes against humanity, war crimes, and ethnic cleansing, to encompass climate change. While a range of options exist that could provide steppingstones towards inclusion, given the actionable powers inherent in the R2P concept, including the use of coercive lethal force, it would seem prudent to only seek a new international consensus on the R2P scope, via the General Assembly, should a series of relevant political and practical changes (e.g. the inclusion of ecocide under the Rome Statute) point in that direction. Next, this section examined several conceptual issues regarding the expansion of the R2P watch list factors: 'past atrocities', 'continuing tensions', 'coping mechanisms', 'external receptiveness' and 'leadership quality'. It was suggested that 'climate vulnerability' is worth considering as the sixth R2P factor based on the significant compatibility between the countries of R2P concern and the countries of climate concern, and the frightening climate projections that will impact most heavily on the regions and human populations already living on the edge of survival.

In summation, Part I traced the long history of atrocities and climatic change (and emissions generation), and the major policy efforts to stop them occurring. It showed that atrocities and emissions are worsening (Chapter 1), and that associated policy efforts have largely failed (Chapter 2). This chapter provided a short history on the recent efforts by Western governments (US, UK, and EU) to integrate climate change into peace and security policy; and opened the discussion about integrating climate change (and climate action) into the Responsibility to Protect – the dominant atrocity prevention and response framework globally.

Part II builds on this discussion by paralleling and integrating climate action policies into the R2P's operational framework (otherwise known as the 'R2P toolbox'), across its prevention (Chapter 4), reaction (Chapter 5), and rebuilding (Chapter 6) phases.

Notes

1 Security Council, 5663 Meeting, S/PV.5663, New York, 14 April 2007.
2 'Climate Security 101', *The Centre for Climate and Security* (accessed 2022). See also, as an example: *Climate Change and International Security*, Paper from the High Representative and the European Commission to the European Council, 14 March 2008.
3 'Climate Security 101', (accessed 2022). See also, Holly Kaufman and Sherri Goodman, 'Climate Change in the U.S. National Security Strategy: History and Recommendations', Briefer, *The Centre for Climate and Security*, 29 June 2021.
4 As suggested by many publications such as Christine Parthemore, John Conger, Francesco Femia, and Kate Guy, 'Confronting Systemic Security Risks: Proposals for The Next U.S. Administration', Policy Briefer, *Council on Strategic Risks*, October 2020.
5 See important report by Dennis Tänzler, Daria Ivleva, and Tobias Hausotter, 'EU Climate Change Diplomacy in Post-Covid-19 World', Study Requested by the European Parliament's Committee on Foreign Affairs, July 2021.
6 'Council Conclusions on EU Climate Diplomacy', Foreign Affairs Council Meeting Brussels, *Council of the European Union*, 18 July 2011; 'Council Conclusions on EU Climate Diplomacy', Foreign Affairs Council Meeting Brussels, *Council of the European Union*, 24 June 2013.
7 Peter Wittig, Permanent Representative Germany, 'Concept Note', Letter dated 1 July 2011 from the Permanent Representative of Germany to the United Nations addressed to the Secretary-General, UN Security Council, 5 July 2011.
8 Ibid.
9 Ban Ki-moon, 'Remarks to the Security Council on the Impact of Climate Change on International Peace and Security', Secretary-General, 20 July 2011.
10 *Statement by the President of the Security Council*, S/PRST/2011/15, Security Council, 20 July 2011. See also, 'Climate Security 101' (accessed 2022).
11 Ibid.
12 'Press Conference on Impact of Climate Change on Marshall Islands', Press Conference, *United Nations*, 15 February 2013.
13 Caitlin Werrell and Francesco Femia, 'UN Security Council Meeting on Climate Change as a Threat Multiplier for Global Security', *The Centre for Climate and Security*, 8 July 2015.

14 'Arria Formula Meeting on Climate Change', *Security Council Report*, 14 February 2013.
15 'Shared Vision, Common Action: A Stronger Europe a Global Strategy for the European Union's Foreign and Security Policy', *European Union*, June 2016. See also, 'Council Conclusions on EU Climate Diplomacy, Foreign Affairs', Council Meeting, Luxembourg, Press Release, 24 June 2013; 'Council Conclusions on Climate Diplomacy', Press Release, *Council of the EU*, 20 July 2015; 'European Climate Diplomacy after COP21 – Council Conclusions', *Council of the EU*, 15 February 2016.
16 'Joint Communication to the European Parliament and the Council: A Strategic Approach to Resilience in the EU's External Action', *European Commission*, 7 June 2017.
17 'Council Conclusions on Climate Diplomacy – Council Conclusions', *European Union*, 26 February 2018, p 3. See also, Oli Brown, Anne Le More, and Julie Raasteen, 'Climate-Fragility Policy Paper: Europe and Climate Security: Is Europe Delivering on Its Rhetoric?', *Adelphi* and *Climate Diplomacy*, July 2020, p 8.
18 'European Parliament Resolution of 3 July 2018 on Climate Diplomacy' (2017/2272(INI)), P8_TA(2018)0280, Climate Diplomacy, *European Parliament*, 2018, paragraph 7.
19 Ibid.
20 Tänzler, Ivleva, and Hausotter, 'EU Climate Change Diplomacy in Post-Covid-19 World', p 20. See also, Brown, Le More, and Raasteen, 'Europe and Climate Security', p 19.
21 'Climate Change: Arria-formula Meeting', *Security Council Report*, 14 December 2017.
22 'Responsibility to Prepare and Prevent', *The Centre for Climate and Security* (accessed 2022).
23 Resolution 2349, Peace and Security in Africa – Lake Chad Basin, S/RES/2349, Security Council, 31 March 2017.
24 On Sudan/Darfur (S/RES/2429, 13 July 2018), Silencing the Guns in Africa (S/RES/2457, 27 February 2019), Democratic Republic of the Congo (S/RES/2502, 19 December 2019, S/RES/2556 18 December 2020), Central African Republic (S/RES/2499 15 November 2019, S/RES/2552 12 November 2020). This analysis combines relevant lists found on 'UN Security Council Resolutions and Presidential Statements Referencing R2P', *Global Centre for R2P*, and 'Climate Security at the UNSC – A Short History', *Climate Diplomacy* and *Adelphi.*
25 On Somalia (S/RES/2408, S/RES/2431, S/RES/2461, S/RES/2472, S/RES/2520, S/RES/2540, S/RES/2568), Mali (S/RES/2423, S/RES/2480, S/RES/2531, S/RES/2584), Sudan (S/RES/2524, S/RES/2579), Central African Republic (S/RES/2448), Iraq (S/RES/2576), and Cyprus (S/RES/2587).
26 IPCC, 2018: *Global Warming of 1.5°C*. An IPCC Special Report on the Impacts of Global Warming of 1.5°C Above Pre-industrial Levels and Related Global Greenhouse Gas Emission Pathways, in the Context of Strengthening the Global Response to the Threat of Climate Change, Sustainable Development, and Efforts to Eradicate Poverty Global Warming of 1.5°C, 2019.
27 Ibid., p 245.
28 'Maintenance of International Peace and Security', Security Council, S/PV.8451, 25 January 2019, p 2.
29 Ibid., p 3.

30 'Summary: UNSC Open Debate on Climate and Security – 24 July 2020', *Climate Diplomacy*, 30 July 2020.
31 'Climate Change and Security', *Security Council Report*, 29 January 2021.
32 'Climate Security at the UNSC' (accessed 2022).
33 'Climate Change and Defence Roadmap', EEAS (2020)1251, *European External Action Service*, 9 November 2020.
34 Louise van Schaik and Akash Ramnath, 'Mission Probable: The EU's Efforts Policy Brief to Green Security and Defence', Policy Brief, *Clingendael Institute*, August 2021.
35 Richard Youngs, 'The EU's Indirect and Defensive Approach to Climate Security', *Carnegie Europe*, 12 July 2021. See also, van Schaik and Ramnath, 'Mission Probable'.
36 Lisa Friedman, 'Biden Introduces His Climate Team', *The New York Times*, 22 December 2022. See also, Francesco Femia and Caitlin Werrell, 'The Biden Administration & Climate Security: Week One', *The Centre for Climate and Security*, 22 January 2021; United States Ambassador to the United Nations, Linda Thomas-Greenfield, 'Remarks at a Joint UN Security Council Stakeout with Germany and Nauru on the United States Joining the UN Group of Friends on Climate and Security', 8 April 2021.
37 'Executive Order (14008) on Tackling the Climate Crisis at Home and Abroad', Presidential Actions, *The White House*, 27 January 2021.
38 Special Presidential Envoy for Climate, John Kerry, 'Remarks on the Urgency of Global Climate Action', *US Department of State*, 20 July 2021.
39 'Fact Sheet: Prioritizing Climate in Foreign Policy and National Security', Statements and Releases, *The White House*, 21 October 2021. See also specific reports, 'DHS Strategic Framework for Addressing Climate change, *US Department of Homeland Security*, 21 October 2021; 'Department of Defense Climate Risk Analysis', *US Department of Defense*, October 2021; 'Climate Change and International Responses Increasing Challenges to US National Security Through 2040', *US National Intelligence Council*, October 2021; 'Report on the Impact of Climate Change on Migration', *The White House*, October 2021. See also for commentary, Shruti Samala, 'All Systems Go: Integrating Climate Security Across the U.S. Government', *New Security Beat*, 18 October 2021.
40 *Report on the Impact of Climate Change on Migration*, A Report by The White House, October 2021.
41 'Concept for an Integrated Approach on Climate Change and Security', EEAS(2021)770, *European External Action Service*, 16 September 2021.
42 'Climate Change and Sustainability Strategic Approach', *UK Ministry of Defence*, 2021.
43 'UK Publishes New Climate-Defence Roadmap', *Centre for Climate and Security*, 7 April 2021.
44 Kate Cox, Anna Knack, Martin Robson, Neil Adger, Pauline Paillé, Jon Freeman, James Black, and Ruth Harris, 'A Changing Climate Exploring the Implications of Climate Change for UK Defence and Security', *RAND Corporation*, 2020. See also, 'UK Publishes New Climate-Defence Roadmap', *Centre for Climate and Security*, 7 April 2021.
45 'Global Britain in a Competitive Age: The Integrated Review of Security, Defence, Development and Foreign Policy', *UK Government*, March 2021.
46 Ibid.

47 Robin Niblett, 'Global Britain in a Divided World Testing the Ambitions of the Integrated Review', *Chatham House*, March 2022, p 18 and 22–23.
48 'Climate Change and Security', *Security Council Report*, 29 January 2021.
49 Barbara Woodward, 'Annex to the Letter Dated 17 February 2021 from the Permanent Representative of the United Kingdom of Great Britain and Northern Ireland to the United Nations addressed to the Secretary-General', 17 February 2021.
50 UK Prime Minister Boris Johnson, 'Climate and Security – Security Council Debate', *YouTube*, 23 February 2021.
51 Security Council, S/2021/990, 13 December 2021.
52 'Climate Change and Security: Vote on a Resolution', *Security Council Report*, 11 December 2021.
53 'Differences Emerge over Appropriate Forum for Discussing Climate Change, as Delegates Hold Debate on Links between Global Crisis, Security', Meetings Coverage and Press Releases, *United Nations*, 23 September 2021.
54 See for example, commentary by John Conger, 'Mainstreaming Climate Security: The FY22 National Defense Authorization Act', *The Centre for Climate and Security*, 3 January 2022; 'US Army releases its Climate Strategy', *US Army Public Affairs*, 8 February 2022; Holly Kaufman and Sherri Goodman, 'Climate Change in the U.S. National Security Strategy: History and Recommendations', Briefer, *The Centre for Climate and Security*, 29 June 2021.
55 '2022 Prologue to the United States Strategy to Prevent Conflict and Promote Stability', *US Department of State*, 1 April 2022. See also, 'Integrating Climate Change into the US Global Fragility Strategy: A New "Prologue"', *The Centre for Climate and Security*, 19 April 2022.
56 '2020 United States Strategy to Prevent Conflict and Promote Stability', US Department of State, US Aid, Department of Defense, Department of the Treasury, December 2020.
57 The report was limited to identifying some the problems and solutions associated with the management of natural resources and environmental assets. See for example, Lauren Herzer Risi, 'New U.S. Global Fragility Strategy Recognizes Environmental Issues as Key to Stability', *New Security Beat*, 14 January 2021.
58 '2022 Prologue to the United States Strategy to Prevent Conflict and Promote Stability', *US Department of State*, 1 April 2022. See also, 'Integrating Climate Change into the US Global Fragility Strategy: A New "Prologue"', *The Centre for Climate and Security*, 19 April 2022.
59 'United States Strategy to Anticipate, Prevent, and Respond to Atrocities', *US State Department*, 15 July 2022; 'From Srebrenica to a Safer Tomorrow: Preventing Future Mass Atrocities Around the World', House of Commons, International Development Committee, 11 October 2022.
60 'UK Approach to Preventing Mass Atrocities', Policy Paper, UK Government, 16 July 2019.
61 'From Srebrenica to a Safer Tomorrow', p 10 and 25, and 54 (witnesses Savita Pawnday, and The Lord Ahmad of Wimbledon), and 55 (submissions number 11, 20, 21, and 26). See also, comments from Inquiry participants in Willian Worley, 'UK Parliament Report Warns Climate Change Likely to Worsen Atrocities', Devex, 17 October 2022.
62 *EU Responsibility to Protect – Atrocity Prevention Toolkit*, European Union External Action, September 2018, p 5.

63 *The Responsibility to Protect*, Report of the International Commission on Intervention and State Sovereignty (Ottawa: International Development Research Centre, 2001), p 33.
64 *2005 World Summit Outcome*, General Assembly Resolution, A/RES/60/1, 24 October 2005, paragraphs 138 and 139.
65 Edward C. Luck, 'Environmental Emergencies and the Responsibility to Protect: A Bridge Too Far?', in Proceedings of the Annual Meeting (American Society of International Law), Vol 103, International Law as Law, 2009, p 33–34.
66 Ban Ki-moon, 'Implementing the Responsibility to Protect', Report of the Secretary-General, A/63/677, General Assembly, 12 January 2009, p 8.
67 Gareth Evans, *The Responsibility to Protect: Ending Mass Atrocity Crimes Once and For All* (Washington, DC: Brookings Institution Press, 2008), p 64–65.
68 Alex J. Bellamy, 'The Responsibility to Protect: A Wide or Narrow Conception?', in Peter Hilpold (Eds.), *The Responsibility to Protect: A New Paradigm of International Law?* (Leiden: Brill Nijhoff, 2015), p 38. See also, Alex J. Bellamy, *The Responsibility to Protect: A Defense* (New York: Oxford University Press, 2015), p 12. See also, Luck, 'Environmental Emergencies and the Responsibility to Protect', p 35.
69 Bellamy, *The Responsibility to Protect: A Wide or Narrow Conception?*, p 38.
70 Lloyd Axworthy and Allan Rock, 'R2P: A New and Unfinished Agenda', *Global Responsibility to Protect*, Vol 1, 2009, p 56.
71 Cyclone Nargis and the Responsibility to Protect', Myanmar/Burma Briefing No 2, *Asia-Pacific Centre for the Responsibility to Protect,* 8 May 2008, p 4.
72 Ibid.
73 Gareth Evans, 'Facing Up to Our Responsibilities', *The Guardian*, 2008.
74 Roberta Cohen, 'Reconciling R2P with IDP Protection', *Global Responsibility to Protect*, Vol 2, No 2, 2010, 15–37.
75 Evans, *The Responsibility to Protect* (Brookings, 2008), p 66.
76 Axworthy and Rock, 'R2P: A New and Unfinished Agenda', p 56.
77 Luck, 'Environmental Emergencies and the Responsibility to Protect', p 35.
78 Joanna Harrington, 'R2P and Natural Disasters', in W. Andy Knight and Frazer Egerton (Eds.), *The Routledge Handbook of the Responsibility to Protect*, p 145. See also the 8 May 2008 report by *The Asia Pacific Centre for R2P* that recognised that the junta could have possibly committed crimes against humanity if they were 'deliberately withholding aid to people who are facing immediate risk of death', but it contended, 'at present, there is no evidence of widespread intent by authorities to cause great suffering' – 'Cyclone Nargis and the Responsibility to Protect', *APCR2P*, p 9.
79 Ashley McLachlan-Bent and John Langmore, 'A Crime Against Humanity?: Implications and Prospects of the Responsibility to Protect in the Wake of Cyclone Nargis', *Global Responsibility to Protect,* Vol 3, No 3, 2011, p 45.
80 North-South relations encompass other key conceptual debates in the R2P literature including non-interference and sovereignty, and northern imperialism and colonialism. See for example, Evans, *The Responsibility to Protect* (Brookings, 2008), p 42; Bellamy, *The Responsibility to Protect: A Defense*, p 84; Gareth Evans, 'State Sovereignty was a Licence to Kill', *International Crisis Group*, 22 May 2008; Luke Glanville, *Sovereignty and the Responsibility to Protect: A New History* (Chicago and London: University of Chicago Press, 2013).
81 Evans, *The Responsibility to Protect* (Brookings, 2008), p 65.

82 Gareth Evans, 'The Responsibility to Protect: Meeting the Challenges', Lecture, Singapore, 5 August 2008. See also Bellamy, *The Responsibility to Protect: A Defense*, chapter entitled 'A Trojan Horse?'.
83 Mely Caballero-Anthony and Belinda Chng, 'Cyclones and Humanitarian Crises: Pushing the Limits of R2P in Southeast Asia', *Global Responsibility to Protect*, Vol 1, No 2, 2009, p 138.
84 Jon Barnett in Tobias Ide, McKenzie F. Johnson, Jon Barnett, Florian Krampe, Philippe Le Billon, Lucile Maertens, Nina von Uexkull, and Irene Vélez-Torres, 'The Future of Environmental Peace and Conflict Research', *Environmental Politics*, 2023, p 7. Barnett was a lead author of 'Human Security' in IPCC, 2014: *Climate Change 2014: Impacts, Adaptation, and Vulnerability. Part A: Global and Sectoral Aspects. Contribution of Working Group II to the Fifth Assessment Report of the Intergovernmental Panel on Climate Change* (Cambridge and New York: Cambridge University Press, 2014), p 755–791.
85 Evans, *The Responsibility to Protect* (Brookings, 2008), p 35.
86 Luck, 'Environmental Emergencies and the Responsibility to Protect', p 35.
87 Evans, *The Responsibility to Protect* (Brookings, 2008), p 65.
88 Linda A. Malone, 'Green Helmets: Eco-Intervention in the Twenty-First Century Unilateral and Multilateral Intervention', Proceedings of the Annual Meeting (American Society of International Law), Vol 103, 2009, p 24.
89 In 1998, the Rome Statute was adopted. It stipulated that the international court, which was launched in 2002, was able to try, convict, and punish individuals for four Crimes Against Peace: genocide, crimes against humanity, war crimes, and the crime of aggression.
90 Robyn Eckersley, 'Ecological Intervention: Prospects and Limits', *Ethics and International Affairs*, Vol 21, No 3, 2007, p 312.
91 Polly Higgins 2013 in Rob White, *Climate Change Criminology* (Bristol: Bristol University Press, 2018), p 20. See also, Polly Higgins, Damien short, and Nigel South, 'Protecting Our Planet: A Proposal for a New Law of Ecocide', *Crime, Law and Social Change*, Vol 59, 2013, p 258. See also, Anja Gauger, Mai Pouye Rabatel-Fernel, Louise Kulbicki, Damien Short, and Polly Higgins, *The Ecocide Project 'Ecocide is the Missing 5th Crime Against Peace'*, Report, University of London, 2012.
92 Rob White, *Climate Change Criminology* (Bristol: Bristol University Press, 2018), p 21. See also, 'Legal Experts Worldwide Draw Up 'Historic' Definition of Ecocide', *The Guardian*, 23 June 2021; 'Make Environmental Damage a War Crime, say Scientists', *The Guardian*, 24 July 2019; 'Stop Military Conflicts from Trashing Environment', Correspondence, *Nature*, 23 July 2019.
93 Shahudha Mohamed, 'Maldives Calls for ICC to Criminalise Ecocide', *The Edition*, December 2019. Sophie Yeo, 'Ecocide: Should Killing Nature Be a Crime?', *BBC*, 6 November 2020.
94 See for example, Jojo Mehta and Julia Jackson, 'To Stop Climate Disaster, Make Ecocide an International Crime. It's the Only Way', *The Guardian*, 24 February 2021.
95 Axworthy and Rock, 'R2P: A New and Unfinished Agenda', p 68.
96 Ibid., p 64 and 67.
97 Eckersley, 'Ecological Intervention: Prospects and Limits', p 304.
98 'Crisis Watch: Tracking Conflict Worldwide', International Crisis Group (accessed 2023).
99 Evans, *The Responsibility to Protect* (Brookings, 2008), p 72.

100 Ibid., p 74–75.
101 Ibid.
102 Ibid.
103 Caitlin E. Werrell and Francesco Femia, 'Climate Change as Threat Multiplier: Understanding the Broader Nature of the Risk', *The Centre for Climate and Security*, 12 February 2015.
104 Gareth Evans, 'Conflict Potential in a World of Climate Change', Address, Berlin, 29 August 2008.
105 Neal Simpson, Nikki Marczak, Alex Bellamy, and Arna Chancellor, 'Climate Change and Atrocity Crimes: The Challenge in The Pacific', *Asia-Pacific Centre for the R2P*, 2020, p 1 and 17.
106 Simon Adams: 'From Global Warming to Genocide Warning: Climate Change and Mass Atrocities', *Global Centre for the R2P,* December 2016; *Mass Atrocities, the Responsibility to Protect and the Future of Human Rights: 'If Not Now, When?'* (Oxon: Routledge, 2021), p 86 and 91. Savita Pawnday in 'From Srebrenica to a Safer Tomorrow: Preventing Future Mass Atrocities Around the World', UK Parliament, 2022.
107 CNA Military Advisory Board, *National Security and the Threat of Climate Change* (Alexandria, VA: CNA Corporation, 2007), p 3, 6, and 44.
108 See for example, 'Climate Change and International Security', Paper from the High Representative and the European Commission to the European Council, March 2008; 'The Quadrennial Defense Review', *US Department of Defense*, 2014, p 8; 'NATO Climate Change and Security Action Plan', Media Release, *NATO*, 14 June 2021; 'African Peace and Security Architecture: APSA Roadmap 2016–2020', *African Union*, 5 December 2015, p 20.
109 For example, climate change is mentioned only twice in the 999-page *Oxford Handbook of the Responsibility to Protect*, See Alex J. Bellamy and Tim Dunne (Eds.), *The Oxford Handbook of the Responsibility to Protect* (New York: Oxford University Press, 2016).
110 Beatrice Mosello, Lukas Rüttinger, and Liesa Sauerhammer, 'The Climate Change Conflict Connection: Current State of the Knowledge', *Adelphi*, 12 November 2019, p 5.
111 Joshua Busby and Nina von Uexkull, 'Climate Shocks and Humanitarian Crises: Which Countries Are Most at Risk?', *Foreign Affairs*, 29 November 2018. Joshua Busby, 'It's Time We Think Beyond "Threat Multiplier" to Address Climate and Security', *New Security Beat*, 21 January 2020.
112 Lukas Rüttinger, Dan Smith, Gerald Stang, Dennis Tänzler, and Janani Vivekananda, 'A New Climate for Peace: Taking Action on Climate and Fragility Risks', An Independent Report Commissioned by the G7 Members, *Adelphi*, 2015, p 6.
113 Elizabeth G. Boulton, 'Plan E: A Grand Strategy for the Twenty-first Century Era of Entangled Security and Hyperthreats', *Journal of Advanced Military Studies*, Vol 13, No 1, 2022, p 92–128.
114 See 'DHS Strategic Framework for Addressing Climate change, *US Department of Homeland Security*, 21 October 2021; 'Department of Defense Climate Risk Analysis', *US Department of Defense*, October 2021; 'Climate Change and International Responses Increasing Challenges to US National Security Through 2040', *US National Intelligence Council*, October 2021; 'Report on the Impact of Climate Change on Migration', *The White House*, October 2021.

115 Rita Floyd, 'Global Climate Security Governance: A Case of Institutional and Ideational Fragmentation', *Conflict, Security & Development*, Vol 15, No 2, 2015, p 119–146. See also, John Vogler, 'Changing Conceptions of Climate and Energy Security in Europe', *Environmental Politics*, Vol 22, No 4, 2013, p 634.
116 See suggestion by Florian Krampe and Malin Mobjörk in 'Responding to Climate-Related Security Risks: Reviewing Regional Organizations in Asia and Africa', *Current Climate Change Reports*, Vol 4, December 2018, p 335.
117 See suggestion about threat-opportunity contrast in Florian Krampe, 'Risk, But Also Opportunity in Climate Fragility and Terror Link', *New Security Beat*, 18 May 2017.
118 'Populations as Risk', *Global Centre for R2P* (accessed 2022).
119 'UN Security Council Resolutions and Presidential Statements Referencing R2P', *Global Centre for R2P* (accessed 2022).
120 'Crisis Watch: Tracking Conflict Worldwide', *International Crisis Group* (accessed 2022).
121 'ND-Gain Country Index: Country Rankings', *Notre Dame Global Adaptation Initiative* (accessed 2021). See also, 'University of Notre Dame Global Adaptation Index: Country Index Technical Report', *Notre Dame Global Adaptation Initiative* November 2015.
122 Ibid.
123 Florian Krampe, 'Why United Nations Peace Operations Cannot Ignore Climate Change', *SIPRI*, 22 February 2021.
124 'Addressing the Impact of Climate Change on Peace and Security', *Department of Political and Peacekeeping Affairs* (accessed 2021).
125 'When Rain Turns to Dust: Understanding and Responding to the Combined Impact of Armed Conflict and the Climate and Environment Crisis on People's Lives', *International Committee of the Red Cross*, 2020.
126 'When Will the World Reach 1.5C of Global Heating?', *The Guardian*, 5 April 2022.
127 IPCC, 2021: 'Summary for Policymakers' in *Climate Change 2021: The Physical Science Basis. Contribution of Working Group I to the Sixth Assessment Report of the Intergovernmental Panel on Climate Change* (Cambridge and New York: Cambridge University Press) p 14 and 22. See also, David Herring, 'Climate Change: Global Temperature Projections', *US Government*, 6 March 2012.
128 Michon Scott, 'What's the Coldest the Earth's Ever Been?', US Government, 18 February 2021.
129 *Closing the Window: Climate Crisis Calls for Rapid Transformation of Societies*, Emissions Gap Report 2022 (Nairobi: UN Environment Program, 2022), p xv and xxi.
130 IPCC, 2021: 'Summary for Policymakers', p 41.
131 See Rüttinger, et al., 'A New Climate for Peace', p 5–6.
132 Camilo Mora and colleagues, 'Broad Threat to Humanity from Cumulative Climate Hazards Intensified by Greenhouse Gas Emissions', *Nature Climate Change*, Vol 8, December 2018, p 1062–1071.
133 'UN Chief Warns of "Breakdown in Trust" With no Deal in Sight at Cop27', *The Guardian*, 18 November 2022; 'Humans Waging "Suicidal War" on Nature – UN Chief António Guterres', *BBC*, 2 December 2020; 'António Guterres on the Climate Crisis: 'We Are Coming to a Point of no Return', *The Guardian*, 11 June 2021; 'We Are Digging our Own Graves': World Leaders' Powerful Words at COP26', *The Guardian*, 2 November 2021; 'UN Chief Says the World is in

'Life-or-Death Struggle' for Survival', *PBS News Hour*, 3 October 2022. See also, "Climate Crisis Hits Worst Case Scenario" Levels – Environment Agency Head', *The Guardian*, 24 February 2021; 'Millions Will Die If World Fails on Climate Promises', *Climate News Network*, 16 February 2021; 'Faster Pace of Climate Change Is "Scary", Former Chief Scientist Says', *BBC*, 16 September 2019; 'Climate Change: "Clear And Unequivocal" Emergency, Say Scientists', *BBC*, 6 November 2019; 'Dangerously Close: Tipping Points May Trigger Climate Cascade', *The Sydney Morning Herald*, 28 November 2019.

134 See Rüttinger, et al., 'A New Climate for Peace', p 6.

Part II

Climate Action and R2P's Operational Framework

4 Climate Action and the Responsibility to Prevent

> Experience has constantly taught us that effective prevention is far less costly in blood and treasure than cure.
>
> (Gareth Evans, *The Responsibility to Protect*, 2008, p 79)

> The benefits of strong and early action far outweigh the economic costs of not acting.
>
> (Nicholas Stern, *The Economics of Climate Change*, 2007, p vi)

Prevention is at the heart of policy efforts to reduce mass atrocities. Atrocity prevention policies are firmly based in the post-1945 establishment of Chapter VI of the UN Charter, which identifies several measures to stop and settle conflict by peaceful means including though 'negotiation, enquiry, mediation, conciliation, arbitration, and judicial settlement'.[1] Prevention policies, and institutional buy-in, surged again in the post-1989 environment. In 1992, the UN Secretary-General Boutros Boutros-Ghali published his report, *An Agenda for Peace*, emphasising the UN's role 'to try through diplomacy to remove the sources of danger before violence results' – and identified a range of policies that decision-makers could consider.[2] Similarly, in early 2001, Secretary-General Kofi Annan published his report, *Prevention of Armed Conflict*, identifying a range of policies and a need 'to move from a culture of reaction to a culture of prevention'.[3] The International Commission on Intervention and State Sovereignty (ICISS) report of September 2001 left no doubt that prevention was at its core, stating in bold that 'prevention is the single most important dimension of the responsibility to protect', and presented a list of strategies and policy options under 'The Responsibility to Prevent'.[4] The 2005 World Summit Outcome Document, adopted unanimously by the 193 Member States of the UN General Assembly, institutionalised R2P's prevention focus in assigning the responsibility to prevent the four atrocity crimes (genocide, ethnic cleansing, war crimes, and crimes against humanity), including their incitement, first to each individual state, with the help of external actors if necessary, and failing this, the wider international community, through the United Nations, using appropriate diplomatic, humanitarian, and other peaceful means (see Chapter 2 for more detail).[5]

DOI: 10.4324/9780367201364-7

Preventing human protection catastrophes also underpins the climate action agenda. In 1992, UN Member States agreed to the text of the UN Framework Convention on Climate Change (UNFCCC) and therein its 'ultimate objective (the) stabilization of greenhouse gas (GHG) concentrations in the atmosphere at a level that would prevent dangerous anthropogenic interference with the climate system'.[6] 'Dangerous', it was later clarified, referred to the threat posed by climate change to human populations.[7] Achieving the 'ultimate objective' of the Convention was stipulated in all major international climate agreements that followed, the Kyoto Protocol in 1997,[8] the Copenhagen Accord in 2009,[9] the Paris Agreement in 2015,[10] and the Glasgow Pact in 2021.[11] To prevent dangerous warming – which was eventually agreed to in Copenhagen as warming that exceeded 2°C above pre-industrial times – the UNFCCC established a variety of emissions mitigation policies such as targets and timetables and climate adaptation policies for instance technology transfer and climate finance (see Chapter 2). More recently, the climate security agenda has sought to develop and implement specific climate-informed human protection policies and strategies such as climate-sensitive mediation and peacebuilding processes.

Early Warning Systems

Early warning systems are the upshot of the need for policymakers to have-to-hand an intelligible analytical tool when making judgements about whether preventive action should be taken. They usually involve identifying the potential causes of a large-scale human protection catastrophe; monitoring and analysing the risk of a human protection situation occurring; translating high-risk situations into early warnings for decision-makers to consider; and converting decision-making outcomes into effective and proportionate responses that mitigates the situation. The section to follow examines in parallel several atrocity and climate early warning systems before presenting some initial thoughts on an integrated approach.

Atrocity Early Warning Systems

Conflict early warning systems in the post-Cold War era can be traced back to Secretary-General Boutros-Ghali's 1992 publication, *An Agenda for Peace*. The publication identified a variety of potential threats to peace and security, including 'environmental threats' and 'natural disasters', and called for closer cooperation across the UN to

> strengthen arrangements in such a manner that information from these sources can be synthesized with political indicators to assess whether a threat to peace exists and to analyse what action might be taken by the United Nations to alleviate it.[12]

Since this time, early warning systems and data-driven models seeking to predict deadly conflict have proliferated, with contributions from national governments, regional organisations, international organisations and their agencies, financial institutions, and civil society groups including universities. Most do not identify climate impacts as a threat, or if they do, do so very sparingly. For example, of the 19 sets of conflict indicators in the African Union Continental Conflict Early Warning System (established in 2002, and abandoned in 2022), only one addressed environmental issues – which was overwhelmingly concerned with natural resources.[13]

R2P-based early warning systems were first articulated in the ICISS report of 2001. The report found that while a variety of conflict early warning systems operated globally (as above), there was limited coordination between them, and most had limited capability to translate early warnings into actionable policy responses.[14] The Commissioners therefore recommended the UN establish a centralised conflict and atrocity early warning system, staffed with a small group of conflict prevention specialists that reported directly to the UN secretary-general.[15] The major follow-up document on R2P, the 2005 World Summit Outcome, specifically prescribed the international community 'support the United Nations in establishing an early warning capability'.[16] In 2009, the UN Office on Genocide Prevention published an early warning risk assessment framework for the sole crime of genocide.[17] In 2014, this framework was replaced with the *Framework of Analysis for Atrocity Crimes* (the Framework), covering all four R2P crimes.[18] The revised Framework was published, as it explains, to better reflect the Office's expanded responsibilities (becoming the 'UN Office on Genocide Prevention and the Responsibility to Protect') as well as 'the recent developments and new research' into the processes that lead to the four crimes.[19]

The Framework's analytical approach can be divided into three planes: background, middle ground, and foreground. The background plane comprises the thematic point that political, social, and economic dynamics should be taken into consideration when making an assessment. But this plane held no specific analytical content itself or organisational capacity for the other planes. The middle ground plane comprises 14 'risk factors' for atrocity crimes; eight factors were considered 'common risk factors' (that is, common to all atrocity crimes), while two were considered 'specific risk factors' (that is, two factors attributable to 'genocide' and 'crimes against humanity' and 'war crimes' – 'ethnic cleansing' was not included because it does not have a distinct legal definition as an international crime). Each 'risk factor' comprises six to 18 'indicators' (also referred to as 'signs' in the Framework), which can be considered the foreground, proximate plane.

To help resolve the issue of translating early warnings into meaningful action, the Framework provides the Special Advisor on the Prevention of Genocide and the Special Advisor on the Responsibility to Protect with opportunities to inform the UN Secretary-General, and through him the UN Security Council, on situations where there is a risk of atrocity crimes and allows the advisors to recommend policy responses.[20] Implementation of policy is largely based on the political will, interests, and calculations of nation-states.

The Framework's limitation is that it does not mention climate change or the environment. In 2020, the Asia-Pacific Centre for the R2P published a report that addressed this deficiency.[21] The report successfully integrated climate-related issues into the Framework's common risk factors (apart from 'capacity to commit atrocity crimes'), and into a select number of indicators.[22] It recommended that the international community, via the UN and INGOs, should

> incorporate a climate change lens into conflict mitigation and atrocity prevention, for example by updating The Framework of Analysis for Atrocity Crimes to include risk factors or indicators that take into account the effect of climate change on atrocity crime risk.[23]

In 2021, the Director of the Global Centre for the R2P, Simon Adams, identified several features of the UN Framework that could possibly accommodate climate impacts, writing that,

> Although the UN's Framework of Analysis for Atrocity Crimes does not specifically address global warming, there are two risk factors that are directly relevant: motives or incentives and triggering factors. Triggering factors could include environmental changes that drastically affect entire societies, including severe drought or natural disaster. Motives or incentives could potentially include a dramatic widening of inequality because of climate change, environmental degradation and the political manipulation of access to depleted natural resources.[24]

Climate-Related Early Warning Systems

The dominant form of early warning system in the climate action space are those that predict climate hazards. Climate hazards, which can be interpreted as proximate and observable 'signs', can be short-term/rapid onset events (e.g. cyclones, floods, heatwaves), and/or long-term/slow onset events (e.g. drought and desertification). These data-driven early warning systems use sophisticated weather radar, satellite technology and geographic information system (GIS) imaging to monitor and predict hazards and assess the risk of disaster. The UN Development Programme (UNDP) uses this approach for its flagship Strengthening Climate Information and Early Warning Systems, as does the UN Environment Programme (UNEP) with its Climwarn Early Warning Systems.[25] These UN-based systems provide risk assessment services in countries across Africa, the Middle East and Asia. Autonomous weather monitoring and prediction stations called, Automatic Weather and Agrometeorological Stations and Automatic Hydrological Stations, have also been set up in a variety of vulnerable communities. In addition, developed and developing countries have established climate risk information-sharing partnerships. For example, the UK MET Office in partnership with UKAid provides a 'multi-hazard early warning service' to the Tanzania Meteorological Agency.[26]

Climate security early warning systems occupy another category, involving the integration of climate and conflict-sensitive programming by linking global climate data (e.g. GIS) with 'hotspot' mapping of conflict risk, and developing inclusive national, regional, and cross-border responses.[27] Several European countries have developed climate security early warning systems. For example, the Dutch government-funded Water, Peace and Security early warning tool combines environmental variables such as rainfall and crop failures with political, economic, and social factors to predict the risk of violent water-related conflicts 'up to a year in advance'. The tool currently focuses on hotspots across Africa, the Middle East and Southeast Asia and 'helps stakeholders to identify and understand water-related security risks and undertake timely, informed and inclusive action for conflict prevention and mitigation'.[28] The German government has been involved in developing several climate security early warning tools including the ClimSec Early Warning System on Human Security in the Sahel Region, and the Global Climate and Security Risk and Foresight Assessment process, which is currently under development in partnership with researchers from the Potsdam Institute for Climate Impact Research, Adelphi's climate security programme and the International Crisis Group, among others.[29]

With regards to moving from analysis to action, climate-related early warning systems that monitor and predict climate hazards tend to be quite basic, for example, 'communicating the risks to individuals, communities, businesses, and governments in fragile countries so they can take timely action' or circulating climate bulletins to vulnerable communities and workers such as farmers so they can make changes.[30] Climate security early warning systems have taken several different approaches. For example, the World Meteorological Organization has sought to strengthen coordination between the UN and national meteorological services and various UN crisis response centres.[31] Similarly, Risk-informed Early Action Partnership, a not-for-profit, aims to enable 'coherence, alignment and complementarity' of existing initiatives, involved in both climate-hazard monitoring and humanitarian response.[32] The German-led Foresight Assessment process takes a slightly different approach to climate security challenges by first identifying a country of concern and then building operational capacity in the country in question long before a human protection situation arises – in short, it takes a pre-emptive approach.[33]

In terms of limitations, in general, climate security early warning systems (as well as climate hazard systems) do not incorporate traditional in-country intelligence gathering into their analysis, rather deferring to information generated by combining real-time weather tracking data with geospatial data (e.g. erosion, water tables, salinization, flood events, forest cover, and population locations) overlaid with hotspot mapping of conflict risks.[34] The key limitation in the present context however is that the climate security agenda is yet to develop an early warning system specifically for climate and atrocity risk assessment (with an effective mechanism to translate early warnings into policy responses).

Towards an Integrated Climate-Atrocity Early Warning System

As suggested earlier, a pathway forward might be to prepare a climate-informed UN Framework of Analysis for Atrocity Crimes. The 2014 version was published prior to the Paris Agreement being struck, the subsequent increase in UN Security Council debates on climate security, and the West's push to integrate climate policy into their foreign and defence policies (see Chapters 2 and 3). Any renovated (or completely new) climate-informed atrocity early warning system could consider using the three planes (background, middle-ground, and foreground) discussed earlier, prescribing equal importance to climate change issues as to traditional atrocity prevention issues (see Table 4.1).

The background plane would establish 'the environment' as a structural dynamic of equal analytical importance to the traditional political, social, and economic dynamics. An extension such as this would demand that analysts

Table 4.1 Example of an Integrated Atrocity and Climate Early Warning Framework

Mass Atrocity Crimes	*Climate-Related Human Catastrophes*
Background Structural Dynamics	
political, economic, social	environmental
Mid-Range Factors	
History of atrocities, continuing tensions, coping mechanisms, receptivity of outside help, quality of leadership	vulnerability to climate change (as in degree of exposure to climate change impacts/hazards)
Foreground Observable Signs	
hate speech against the vulnerable	hotter and dryer conditions
growing repression	diminishing water resources, water scarcity
human rights abuses	intensifying storms and cyclones
historical grievances	rising sea levels
Acute inequality	habitat loss
Poor governance	migration
Recent social traumas	food scarcity and stress
Prejudice (e.g. based on religious or ethnic grounds) towards health and education	malnutrition
Rapid social, economic, and political dislocation	related impacts on agriculture
Colonial occupation, war, revolution	conflict

examine the broader environmental history of a given situation (e.g. past environmental/climate-induced tensions). This should not be controversial, indeed Evans himself discusses the importance of understanding background dynamics, as does the UN Atrocity Framework and its climate rendering by the Asia-Pacific Centre for the R2P.[35] The mid-level plane would establish 'vulnerability to climate change' as the sixth R2P 'factor'. As discussed in Chapter 3, Evans identifies five factors that decision-makers should consider when making the call whether a country is one of 'R2P concern': history of atrocities, continuing tensions, availability of coping mechanisms, receptivity of external influence, and quality of leadership.[36] Extending the R2P factors to include 'climate vulnerability' helps modernise the concept by acknowledging that climate and atrocity hotspots overlap and the enormity of the climate problem. The foreground plane would establish a variety of observable 'signs' (or 'indicators') of climate impacts to be measured alongside the many early warning signs of atrocities such as increased hate speech directed towards minorities. Drawing on insights from both R2P and climate early warning systems, each plane could be informed by a combination of data-driven models and case-by-case field-based intelligence gathering and qualitative analysis, working in collaboration with international, regional, national, and subnational bodies.[37] Developing such a system would provide the climate agenda with an atrocity-focused early warning system (which draws on existing ideas from the atrocity prevention agenda); and provide the atrocity prevention agenda with a climate-informed atrocity early warning system (which integrates technologies and approaches found in the climate action agenda).

In 2018, the Climate Security Mechanism (CSM) was created in the UN Department of Political and Peacebuilding Affairs (see Chapter 3). The CSM works across UN agencies on climate security issues, including on issues related to climate-informed conflict early warning systems; and therefore seems well placed to work with UN atrocity prevention agencies to translate warnings into effective preventive policy action.[38]

The remainder of this chapter parallels the suite of policies and ideas found in the R2P's prevention 'toolbox' (as sequenced and expressed in Evans 2008)[39] with those found in the climate action agenda, before, on occasions, identifying possible R2P-climate linkages and applied policy actions. It commences with the R2P's political and diplomatic options.

Political and Diplomatic Strategies

Structural and Longer-Term Political Measures

Support Good Governance

Supporting representative, responsive, accountable, and capable governance (or 'good governance') is at the heart of the responsibility to prevent.[40] Very often, violent conflict is driven by failures of basic governance including the long-term misrule and corruption by political and economic elites and the associated socioeconomic and institutional decay that it brings.[41] Achieving good governance involves individual states themselves and the broader international community playing a mutually supportive role in developing the institutions and processes – legislative, executive, and judicial – of effective government, combined with a willingness by individual states to accept a strong civil society as well as basic political and civil rights.[42]

The challenge of 'climate governance' stems from three interrelated factors: the multiple scales and levels of political decision-making involved; the need for coordination and collaboration between state and non-state actors; and the deeply embedded nature of GHG emissions in everyday processes of production and consumption.[43] These factors become especially challenging in states with already fragile and fragmented governance structures and limited capacity (or desire) for international engagement; where civil society actors are marginalised or worse; and/or where regime elites are entrenched in emissions intensive industries such as manufacturing and mining. Despite the challenges, efforts should be made to integrate climate adaptation and mitigation considerations across all government policy domains and objectives in fragile states, including budgeting. This process has been termed 'climate mainstreaming' and has become increasingly popular in the West.[44]

There are four climate mainstreaming options that graduate in strength. First, and the weakest option, is to establish climate mainstreaming as the central priority in the governments' climate change plans. This approach elevates mainstreaming in government decision-making but allows significant scope to exclude climate considerations in broader public policy decisions. Second, include climate mainstreaming in the governments' central strategic policy document. This option sends a stronger signal to those agencies and decision-makers outside the core climate action sectors (energy, transport, agriculture, and heavy industry) that there is a need to consider climate change. Third, legislate a mainstreaming approach in national climate change and/or environmental protection act/s. This approach formalises a requirement for decision-makers to consider climate action across ministries and agencies. Fourth, and the strongest option, involves including climate mainstreaming as a constitutional provision, which serves to deeply embed good climate governance in the functioning and values of the state.[45]

Given that conflict-affected and fragile countries are often the most vulnerable to the impacts of climate change, it would seem prudent to initiate a process that encourages government decision-makers to consider both climate-related risks and opportunities in all major decisions. Western governments could assist by supporting 'good climate governance' programmes, either by creating new programmes or by adjusting their many existing governance programmes.[46] A first step in any such programme would be to assess the extent to which climate mainstreaming already operates in the target state.[47]

Encourage Membership of International Organisations and Regimes

An accompanying measure that can contribute to building a preventive structural safeguard against states committing mass atrocities is to encourage states to join and participate in international organisations and regimes (global, regional, bilateral treaties and arrangements).[48] Involvement in these forums helps build trust, frameworks for dialogue and cooperation, as well as 'imposes a multitude of good governance disciplines not just on existing members but aspiring ones'.[49] For Evans, the EU is 'perhaps the most effective structural conflict prevention measure in history'.[50]

The same could be said for the EU's impact on its members in terms of climate action. The EU has long been a global leader on climate mitigation policies both within its own jurisdiction and by encouraging states beyond Europe to do more. For example, the EU Emissions Trading System (EU ETS) was the first large-scale carbon pricing mechanism in the world. Established in 2005, at present the EU ETS places a limit on the overall emissions of approximately 11,000 power stations and manufacturing plants covering all 28 EU Member States plus Iceland, Norway, and Liechtenstein.[51] The EU Member States are also required to comply with a range of other climate-related disciplines such as limits on transport emissions and improvements in land use practices.[52] Since 2005, carbon pricing schemes have proliferated globally. According to the World Bank, as of 2020 'some 40 countries and more than 20 cities, states and provinces already use carbon pricing mechanisms, with more planning to implement them in the future'.[53] GHG mitigation regimes, such as emissions trading schemes, protect vulnerable populations by limiting warming and therefore reducing the severity of the physical impacts over the longer term. NATO membership also has a domestic disciplinary function. Member countries are required to use new military equipment and infrastructure that, first, ensures NATO troops can effectively operate in tougher climatic conditions – such as in response to 'climate-related disasters'; and second, helps to cut emissions from military activities – such as by using biofuels, developing hybrid vehicles and improving energy efficiency.[54] The UK, US, and the EU have numerous bilateral arrangements in place with developing countries to provide finance, technology, and administrative support, along with other assistance, to help vulnerable populations adapt to climate change as well as to encourage low carbon growth.

Direct Political Measures

Enact Preventive Diplomacy

All diplomacy is in a sense preventive with core tasks involving the management of differences, addressing potential disputes, and at a senior level, resolving real conflicts before they become violent.[55] But since the 1990s, the concept of 'preventive diplomacy' has gradually become a distinct approach with its own set of tools to help stop wars before they get started. These tools include: the direct 'good offices' role of the UN secretary-general and his staff, fact-finding missions, eminent persons commissions, conciliation and mediation, friends' groups, and non-official 'second-track' dialogue.[56]

To take one example, the UN DPPA is the main operational arm for the conduct of the secretary-general's 'good offices'.[57] The DPPA's central mission is to sustain peace and prevent conflict. To achieve this, it develops and deploys a variety of conflict prevention and resolution strategies usually in collaboration with other UN agencies, and regional and in-country partners. Since his appointment in 2017, UN Secretary-General António Guterres has repeatedly warned that climate change threatens peace and security.[58] This view is reflected by the DPPA Under-Secretary-General, Rosemary DiCarlo, who believes that 'climate change undermines our core objectives of conflict prevention and sustaining peace'.[59] To keep pace with the threat posed by climate change, the DPPA has increasingly centralised climate change in its operations. For example, and most prominently, in 2018, the DPPA, jointly with the UNDP and the UNEP, established the CSM – with the UN Department of Peace Operations joining in 2021. The CSM aims to advance a 'common approach to the analysis of climate-related security risks and shape integrated and timely responses' by drawing on the expertise from more than 20 UN entities, and a variety of think tanks, academic institutions and practitioners from regional organisations.[60] A practical achievement in this respect was the development of a 'climate security toolbox'.[61] The tools include climate-informed security risk assessments, conflict-sensitive climate adaptation programmes, climate-sensitive peacebuilding initiatives, and climate-informed mediation.[62]

Mediation is a key duty for the DPPA, and it offers a range of support to countries and conflictual parties through its Mediation Support Unit (MSU) – embedded in which is a digital mediation toolkit and standby team of meditation experts deployable within 72 hours.[63] Over the years, the DPPA, sometimes partnering with other UN agencies, has published a range of substantive mediation guides for use in the field. While some include climate change, most do not.[64] In the guides that do, for example, the joint DPPA–UNEP 2015 publication, *Natural Resources and Conflict Mediation Guide: A Guide for Practitioners*, climate change is generally understood as an issue that mediators and the parties to the mediation must grapple with alongside managing tensions over natural resources such a water allocation, fertile land, forests, and even oil and minerals.[65] Against this backdrop, the CSM identified a specific need for the DPPA to develop a fully integrated climate-informed approach to mediation. In 2022, this came to pass with the publication, *The Implications of Climate Change for Mediation and Peace Processes*.[66] The publication details a variety of practical suggestions that conflict mediators should consider – such as undertaking climate-informed risk assessments on environmental assets and leveraging climate finance frameworks – when negotiating a peace agreement between conflicting parties. Western governments have begun to integrate climate change issues into their on-the-ground preventive diplomacy, including mediation.[67] However, as key partners of the CSM, they also have an opportunity to help develop and implement multilateral programmes.[68]

Threat of Political Sanctions

The R2P toolbox insists that only if direct preventive diplomacy should fail, then, and only then, should a harsher range of options come into play.[69] Political sanctions are the next option in the graduating list of options – the threat of which may be enough to elicit the desired behavioural change in the target state and should be used first – and includes such measures as diplomatic isolation, membership suspension from an organisation, travel and asset restrictions on targeted persons, and 'naming and shaming'.[70]

The UK government imposes political sanctions regimes (that is, a collection of sanctions, namely, asset freezes, travel bans, and arms embargoes) based on geography (that is, related to a particular country or region), or theme (that is, related to a particular issue), and/or, on individuals.[71] The UK's thematic sanctions regimes cover issues related to chemical weapons, cyber activity, terrorist groups, and corruption, among others. These regimes leave little scope to integrate climate-related issues. However, in 2020, the UK added a *Global Human Rights* sanctions regime to the list, which is 'aimed at deterring and providing accountability for serious violations of an individual's human rights' encompassing their 'right to life', right not to be subjected to cruel or degrading treatment, and right to be free from slavery.[72] The regime imposes political sanctions on more than 50 individuals involved in large-scale abuses, such as those in command of organising massacres in Myanmar in 2017, through to one-off events such as those involved in the killing of journalist Jamal Khashoggi in 2018.

It is foreseeable that this regime could be extended to include climate-related human protection issues based on three general observations. First, the stated purpose of the UK's sanctions regimes is to support the UK's 'foreign policy and national security objectives, as well as maintaining international peace and security'.[73] As we have seen, climate change is increasingly becoming a standout feature in the UK's foreign and defence policy documents, and successive UK governments have led global efforts for climate change to be addressed through the UN Security Council based on a long-held position that it threatens international peace and security (see Chapters 2 and 3). Second, the relationship between human rights and climate change has become well-established in international policy discourse. For example, UN High Commissioner for Human Rights, Michelle Bachelet, regularly identifies the threat climate change poses to human rights and based on this, advocates for developed and developing countries to strengthen their climate actions.[74] Third, the UK's human rights sanctions regime provides a promising entry point and a possible pathway to explicitly include climate change through the 'right to life' category. The UN human rights office makes the prominent point, which is consistent with the climate science, that 'The human implications of currently projected levels of global heating are catastrophic'.[75] And as we have also seen, the UN Secretary-General, and many Western leaders, have routinely argued that the impacts of climate change as projected threaten vast swathes of the global human population. Taken together, these three observations may provide some degree of impetus for revisiting the relationship between the UK's human rights sanctions regime and climate change.

It should also be noted that the US and EU have human rights sanctions regimes in place. In 2012, the US established the Global Magnitsky Act, named after the Ukrainian-born Russian lawyer, Sergei Magnitsky, who was killed in detention.[76] The Act allows the US government to sanction human rights violators, freeze their assets, and ban them from entering the country. In 2020, the EU established a human rights sanctions regime.[77]

Economic and Social Strategies

Structural and Longer-Term Economic Measures

Encourage Economic Development

Economic deprivation and hopeless poverty can drive individuals towards organisations that use violence as a political tool or fuel grievances and political tensions between groups and communities. By contrast, economic development – and more so, its associated benefits such as employment and an income – helps set the political conditions that reduce the likelihood of mass atrocities being committed.[78] Policy measures that encourage economic development include those that reduce inequality and equalise opportunities across the economy, eliminate corruption and increase government efficiency, deliver better terms-of-trade and improve regional economic integration; special assistance programmes for specific projects such as transport, infrastructure, and improved natural resource management; and those that mobilise external financial support such as direct budgetary support or debt relief in order to encourage investment.[79]

Climate change has the potential to negatively affect economic development (and poverty alleviation, and therefore human rights and security) in two keys ways: first, by making it harder, if not impossible, to achieve future development goals (e.g. the Sustainable Development Goals, consisting of 17 interlinked goals to be achieved by 2030); and second, by reversing the decades of development gains made across the developing world, pushing millions back into poverty. These twin concerns have been voiced repeatedly over many years by representatives from across the UN system including The World Bank, World Health Organization, Food and Agriculture Organization, United Nations International Children's Emergency Fund, and the DPPA, UNFCCC, UNEP, and UNDP.[80] The extent to which economic development is slowed, and perhaps even undone, is dependent on how warm the world gets. For example, the IPCC's *Special Report on Global Warming of 1.5°C*, released in October

2018, found that 'limiting global warming to 1.5°C rather than 2°C above pre-industrial levels would make it markedly easier to achieve many aspects of sustainable development, with greater potential to eradicate poverty and reduce inequalities'.[81] This finding was reflected 10 months later in the UN's annual report on the Sustainable Development Goals.[82]

Three key domains of Western policy action could be explored to shore up economic development in climate-exposed fragile states. First, special development funding could be allocated towards climate mitigation projects such as urban public transport networks, low-pollution energy production, and the use of innovate low-emission materials such as green metals (e.g. steel and aluminium) and concrete as they become more economical.[83] However, the 'green growth' mantra, which was particularly strong during the Covid-19 pandemic, has since receded as fossil fuel production and consumption again took precedent.[84] Second, climate adaptation funding can assist development through public works programmes such as creating natural sea barriers, water conservation and climate-smart agriculture programmes, and Disaster Risk Reduction (DRR) programmes to name a few. However, developed countries have not always made good on their promises to deliver climate finance, most notoriously with the failure to provide $100 billion per year in climate finance to developing countries by 2020 as agreed at COP15 in Copenhagen in 2009.[85] Third, economic development in fragile states could be generated through 'compensation' programmes.[86] The Twenty-Seventh Conference of the Parties (COP 27) held in Egypt in November 2022 focused on this issue. The Parties agreed to establish a 'Loss and Damage' fund involving developed countries compensating developing countries for the effects of climate change especially droughts, floods, cyclones, and other disasters.[87] But it remains unclear which countries will donate, who the recipients will be, and the rules around allocating funds.[88]

Support Social Programmes

International support to help establish formal education programmes that teach tolerance, cooperation, and the development of positive attitudes between members of previously hostile or suspicious communities, is an important item on any agenda seeking to address the underlying causes of mass atrocities.[89] Another way that foreign governments can help dispel the fear, mistrust, and prejudice so often at the heart of intractable conflicts is to support 'community peacebuilding initiatives' that involve international civil society organisations working with local actors at the grassroots level.[90] Community peacebuilding initiatives operate across the developing world, with a particular operational focus on conflict-affected and fragile countries. They are community-led processes to peacebuilding that can complement state institutions that are often weak in capacity or political will. Some community peacebuilding initiatives focus on bringing young people and teenagers together, others focus on women in peacebuilding, while others on children. They use different techniques to create a forum for respectful interactions and dialogue such as the use of various multimedia and social networking platforms, live drama performances and music, sport, and facilitating debates and discussions in neutral settings and offer training opportunities.[91] The interactions and shared learnings are designed to build trust and cooperation on issues such as human rights, economic development, democracy, human health, and or the environment in the context of preventing deadly violence between groups. Some community peacebuilding initiatives such as the Search for Common Ground are starting to include discussions about the relationship between 'new conflict dynamics' such as climate change and peacebuilding.[92]

Climate adaptation programmes are increasingly being delivered at the local scale though, for example, community-based adaptation (CBA) initiatives.[93] CBA initiatives are small-scale grassroots-driven programmes that have synergies with the broader development aspirations of the country in question. CBA initiatives have grown in popularity in developing countries because they integrate local knowledge, capabilities, priorities, and context into the development and implementation of a project, which stands in contrast to the traditional top-down approaches developed and administered from abroad.[94] Practical CBA initiatives are vast and can include for example, trialling drought-resistant seed varieties and establishing plant nurseries; constructing livestock shelters and integrating new climate-proof breeds into existing herds; the provision and installation of rainwater tanks for water harvesting and carving swales into landscapes for water conversation and soil improvement; constructing dykes, dams, and sea walls and/or relocating and resettling entire communities; measures to control insect-borne disease; and the provision of Disaster Risk Reduction handbooks and age-appropriate climate impact awareness exercises such as movie nights and animations. Western governments contribute to CBA initiatives indirectly and directly. For example, the US, UK, and several EU countries contribute funding to the Global Environment Facility, which funds a variety of CBA initiatives through its Small Grants Programme.[95] More directly, UKAid provides funding to INGOs, such as CARE International, to establish CBA projects in select countries in Africa, Asia and the Pacific, Latin America, and the Caribbean.[96] While USAid has funded CBA initiatives across the Pacific and elsewhere.[97] By initiating and facilitating dialogue on shared values and interests and by providing the skills and tools communities collectively need to plan, make decisions, and implement adaptation strategies, CBA initiatives can serve to unite a disparate community around a common objective.

Direct Economic Measures

Aid Conditionality

Direct forms of aid conditionality involve threatening, suspending or withdrawing development assistance as to elicit a particular policy response from the recipient – for example, the holding of proper elections or the cessation of direct human rights abuses.[98] Used in this way, aid removal becomes the functional equivalent of a broad-based economic sanction.[99] Climate-related development aid by and large, refers to climate finance such as government grants for projects in developing countries, but also, more controversially, government or public-private sector loans. It can also refer to the transfer of technology and equipment, technical skills and expertise, and data and assessment tools. Climate aid can be delivered though international channels such as the UN Green Climate Fund, Global Environment Facility, or Adaptation Fund, or regional bodies such as the African Development Fund or Asian Development Fund, and/or through bilateral arrangements, or in collaboration with the private sector, to name a few options.[100] Despite failing to deliver on some major international climate finance initiatives, collective climate-focused development assistance (broadly defined climate finance, plus other material assistance) by Western countries still runs into many hundreds of billions of dollars per year.[101] Indeed, the US and UK are the first and third largest foreign aid donors in the world, respectively – and both have announced that climate action is central to their overseas development aid.[102]

Direct climate aid conditionality in the style envisaged by the R2P framework has been used by the US and UK with significant affect. For example, US President Barack Obama recalls in his memoir how he managed to secure the Copenhagen Accord climate agreement in the final hours of the COP15 negotiations:

> If I leave this room without an agreement [President Obama told Premier Wen of China, Prime Minister Singh of India, and Presidents Zuma and Lula of South Africa and Brazil respectively], then my first stop is the hall downstairs, where all the international press is waiting for news. And I am going to tell them that I was prepared to commit to a big reduction in our greenhouse gases, and billions of dollars in new assistance, and that each of you decided it was better to do nothing. I am going to say the same thing to all the poor countries that stood to benefit from that new money. And to all the people in your own countries that stand to suffer the most from climate change.[103]

Similarly, the UK has used aid conditionality as a tool to secure support for alliance-building activities in the UNFCCC process.[104] The removal of such climate aid (including prospectively as was Obama's tactic), in practice, may mean removing support to subsistence farmers for climate-resilient crops that can grow in hotter, drier conditions, or the implementation of an early warning system in an area vulnerable to flooding, which can have the undesired effect of increasing human suffering in already vulnerable communities (while for the most part, economic and political elites remain insulated from any significant impacts).[105]

Threaten or Apply Economic Sanctions

Direct broad-based economic sanctions – or at least the threat of them – involve measures such as comprehensive or selective trade embargoes, the freezing or dissolution of trade agreements, the withdrawal of investments, or the withdrawal of World Bank or IMF support. Any country or group of countries can threaten or apply economic sanctions without running afoul of international law. However, it is generally agreed that economic sanctions are a very significant step in a graduated process of international pressure falling short of war and should therefore attract serious assessment of the negatives and positives involved before proceeding.[106]

Western governments impose trade restrictions on a long list of countries, which, for instance, involve placing controls on the export or import (transfer) of technology, services, and/or personnel, among other things.[107] Technology transfer has been an important topic in the international climate negotiations since the 1992 Convention in which the parties committed themselves under Article 4.5 to 'promote and cooperate in the development, application, and diffusion, including transfer, of technologies'.[108] Technology transfers usually involve North-to-South transfer, since technologies have so far been mostly developed in highly industrialised countries.

UK, EU, and US climate action programmes facilitate direct and indirect (e.g. via the UNFCCC's 'Technology Mechanism') access to mitigation technology such as large- and small-scale renewable electricity sources, low-emission transport systems and fuels, emerging low-emission options for heavy industry production processes (steel and cement production for example), and consumer options such as energy-efficient appliances; as well as a range of adaptation technologies such as those that promote efficient water use.[109] They also provide regular information through training and support personnel. The transfer of Western climate mitigation and adaptation technology and associated personnel could be restricted to countries that are marching towards egregious human protection violations. Such a suspension may be particularly effective when focused on mitigation activities because energy, transport, and heavy industry sectors in fragile states are generally owned and operated by regime officials and associates. By contrast, restricting support for adaptation technologies may cause unnecessary direct harm to an already highly vulnerable population.

One specific option that would be palatable at home and potentially impactful on regime officials abroad, might be to follow the UK's lead and begin to withdraw from investing in fossil fuel projects overseas.[110] A less desirable option because of its potential to impact on the population at large, would be to threaten or withdraw funding to multilateral climate-related funds, which have been established to assist developing countries prepare for climate disasters and develop low-emission economies.[111] However, this is not without precedent. For example, in 2018 while announcing plans to exit the Paris Agreement, US President Trump also announced that the US would cancel its remaining $2 billion contribution to the Green Climate Fund.[112]

Promise or Deliver Economic Incentives

The promise or delivery of economic incentives can also be used to modify government behaviour or policies in a target country. Economic incentives (sometimes called 'positive inducements') can take such forms as favourable taxation regimes such as tariff reductions, new investment, access to technology, and the lifting of existing sanctions.[113] In many instances, economic incentives can be more effective than sanctions simply because incentives tend to foster goodwill and cooperation, whereas sanctions create hostility and separation.[114] The effectiveness of economic incentives (and sanctions for that matter) can be improved if the following three elements are considered.[115] First, the goal of the incentives should be clearly articulated and understood. Second, the incentives should be coordinated with allies and partners to ensure the target country cannot obtain the incentive (e.g. investment or technology) from another source. Third, they should be sensitive to the policies and politics of the recipient country, including the government's relationship with the private sector.

Western governments have a range of climate-related positive inducements that they can employ, individually or in combination, to encourage governments in target countries to adopt climate action strategies to reduce emissions and protect their population from direct and indirect climate-related harm such as drought, loss of livelihoods, and possibly even conflict. First, in terms of new investments and trade, western governments can transfer and fund new mitigation and adaptation technologies, and expertise, through a range of bilateral and multilateral channels and incentivise private investment overseas. Second, with regards to modernising infrastructure, governments can grant loans to the target government to renovate old and build new energy and transport systems. Loans can be under preferential conditions for example meeting GHG reduction targets or the targeted roll-out of adaptation strategies. Third, as to financial cooperation, donor and recipient government ministers can sign a declaration on financial cooperation and agree to open a currency swap line to improve the stability of the target country's finances – incentivising foreign and domestic private investment in climate action activities in the target country. Fourth, in relation to worker and training exchanges, donor and recipient countries can establish employment and training exchanges on the technical, legal, and social issues relating to climate mitigation and adaptation activities. Fifth, regarding development aid, governments can promise or deliver increased development assistance to commence or progress tailored climate action programmes in the target country.[116]

Constitutional and Legal Strategies

Structural and Longer-Term Legal Measures

Develop Constitutional Structures

For most populations, a sense of security begins with confidence that they live under fair constitutional structures not inherently hostile to their interests.[117] Conversely, problematic constitutions or the inability to agree on new constitutional arrangements can explain many cases of violent conflict the world over.[118] Constitution making includes both the process of making a new constitution or amending an existing constitution as well as the substantive decisions about the design, form, and content of the new or amended constitution.[119] Because of the enormous domestic sensitivities involved with constitution making however there are limits to what foreigners can contribute apart from offering creative technical assistance on possible power-and-wealth-sharing solutions, drawing on a vast array of different models already in existence.[120]

Constitutions that include climate change provisions can be the catalyst for elevated climate action policies and legislation, and therefore, can act as a structural conflict prevention mechanism. Two prominent models of successful constitution making involving climate change and environmental protection in peace processes can be found in South Africa in 1994 and Kenya in 2010. Both constitutions stipulate that all citizens have a right to a clean and healthy environment (or, at least an environment not detrimental to their health and well-being). For example, South Africa's National Constitution – signed into force by Nelson Mandela in 1994 – enshrines a citizens' right 'to an environment that is not harmful to their health or well-being' and 'to have the environment protected, for the benefit of present and future generations, through reasonable legislative and other measures'.[121] These provisions provide the foundation for South Africa's national climate policy settings, which includes a nationwide carbon tax, along with a raft of regional initiatives.[122] The Constitution of Kenya, adopted in 2010, similarly stipulates that 'every person has the right to a clean and healthy environment' that includes the 'right to have the environment protected for the benefit of present and future generations through legislative and other measures'.[123] This stipulation has spurred a raft of continent-leading national and subnational climate initiatives, such as Kenya's National Climate Change Action Plan of 2013, which 'commits all government ministries, departments, and agencies to mainstream climate change across their functions and processes'.[124]

To ensure that climate change is appropriately incorporated in national constitutions – which could involve bold environmental protection provisions as above, through to advice about resource and land allocation provisions – efforts should be made to include climate constitution specialists in constitution making process as early as possible. Several UN agencies offer technical assistance to constitutional reform processes if requested by a national government or mandated by a UN legislative body.[125] Assistance can include advice on the rule of law, gender, and human rights; legal and technical capacity building; financial and administrative support; as well as political support through the secretary-general's 'good offices'.[126] The DPPA's MSU includes constitution making specialists[127]and advises its mediators to 'facilitate the integration of climate change considerations into subsequent or parallel processes, such as a national dialogue or constitutional reform process'.[128]

Promote Respect for Human Rights

Promoting respect for human rights in a target country can help prevent violent conflict.[129] This can be achieved through funding and implementing programmes that help countries and communities develop and/or expand basic protections on freedom of expression, association, religious practice, and the use of language, as well as non-discrimination in employment.[130] The Universal Declaration of Human Rights (1948) does not mention environmental rights. However, the notion that human rights and climate change are interrelated has gained significant ground since. For example, in 2020, Secretary-General António Guterres told the UN Human Rights Council that 'The climate crisis is the biggest threat to our survival as a species and is already threatening human rights around the world'.[131] The UN Office of the High Commissioner for Human Rights (OHCHR) reflects these views, explaining that 'climate change threatens the effective enjoyment of a range of human rights including those to life, water and sanitation, food, health, housing, self-determination, culture and development'; and therefore, 'states have human rights obligations to prevent the foreseeable impacts of climate change, and ensure that those affected by it have access to effective remedies and means of adaptation'.[132] In July 2022, the UN General Assembly voted overwhelmingly to declare the ability to live in a 'clean, healthy and sustainable environment' a universal human right.[133]

To effectively promote respect for OHCHR's list of climate-related human rights requires a sustained multifaceted effort, working closely with local champions. Western countries could provide funding and support to a variety of activities through their human rights budgets and programmes. Activities could include helping to create local communities capable of investigating and defending climate-related human rights. Such assistance could be extended to providing support to track down violators and set-up appropriate judicial bodies to prosecute them.[134] Global NGOs traditionally in the human rights domain such as Amnesty International and Human Rights Watch are also increasingly involved in climate-related human rights advocacy.[135] They join a suite of NGOs devoted to campaigning specifically on the human rights-climate action nexus such as the Climate Justice Foundation – established in 2010 by Former President of Ireland and UN Human Rights Commissioner, Mary Robinson.[136] But there is only so much foreigners can do. Changes in human rights, above all, must be driven, and seen to be driven, from within.[137]

Promote Respect for the Rule of Law

Promoting respect for the rule of law can help reduce the systematic manipulation, disregard, and/or abuse of the law by government officials and the population at large.[138] Options that can help change the legal landscape in a target country include directing funds towards the proper resourcing of courts that are staffed with effective, honest, and independent judges and legal professionals; supporting access to affordable legal remedies for the population; and helping to develop clear and reasonable legal rules not least for those governing commercial transactions.[139]

Climate-related national and subnational laws, such as those prescribing low carbon transitions in the electricity, transport, and land-use sectors, are rapidly proliferating across the developed and developing world, including in countries at risk of atrocities, which by and large have some form of climate-related laws in place, with many having multiple laws covering general environmental protection, electricity production, forestry, and climate hazards.[140] By contrast, only in developed countries is litigation proliferating against governments and their agencies that are deemed to have provided inadequate climate laws and protections for citizens. For example, in the US there have been more than 1500 climate-related litigation cases, compared to zero in the countries that the Global Centre for the R2P identify as at risk of atrocities.[141] Western governments could develop programmes that provide vulnerable populations in conflict-affected countries with access to legal remedies to rectify environmental injustices in their communities. Inspiration could be drawn from the Biden Administration's creation of the Environmental and Climate Justice Division within the US Department of Justice, which has been given greater powers to redress the 'disproportionate harm' that poorer communities across the US have faced from 'climate change and environmental contaminants' for decades.[142] Western countries could also open-up opportunities in universities located in donor and recipient countries to train legal professionals on emerging regulations and contractual transactions relevant to the low-pollution transition.

Eradication of Corruption

Eradicating corruption is the single most important rule of law issue.[143] Corruption can restrict economic growth, undermine the effectiveness of foreign aid, destroy trust in government and public institutions and officials, and in some instances, tip countries into state failure and precipitate conflict.[144] To effectively stamp out corruption requires the involvement of the government, the judiciary, and civil society in the target country, together with support from the international community including businesses and NGOs – in other words, an integrated approach.[145] Countries rich in environmental assets are particularly vulnerable to corruption.[146] It is well established that widespread corruption exists within state structures and through collusion between police and organised crime in the wildlife trade, poaching, mining and the production of oil and coal, as well as land-clearing and agricultural activities.

Climate action programmes are not immune from corrupt practices. Transparency International (TI) – an international NGO that employs an integrated approach to tackle corruption[147] – explains that billions of dollars in climate finances are now flowing annually from developed to developing countries (they estimate roughly $600 billion from national budgets alone) through 'new, uncoordinated and unmonitored channels, with grey areas and loopholes in regulations', which create tempting opportunities for corruption.[148] These funds are used in part for mitigation projects such as those related to sustainable transportation, the development of renewable energy sources, as well as the protection of forests – all of which are embedded in industries (construction, energy, and forestry) that have particularly high levels of corruption risks, with established pathways to siphon off incoming financial flows into private accounts, explain TI researchers.[149] Climate funds are also used for adaptation projects, with inherent dangers of corruption as well.[150] TI rightly claim that 'we can't tackle the climate crisis without tackling corruption'.[151]

The corruption-climate-atrocity nexus becomes a little clearer when comparing Transparency International's 'Corruption Perception Index' with the University of Notre Dame's 'Countries Vulnerable to Climate Change' index and the Global Center for the R2P's 'Populations at Risk' of atrocities index. This comparison shows that the countries identified as high or highly corrupt are the same as those identified as highly vulnerable to climate change which are very closely matched with those identified as at risk of atrocities.[152] The US, UK, Germany, and France are among the largest contributors to international climate funds.[153] To ensure that climate finance reaches those vulnerable people who most need it, donor governments could develop and implement integrated corruption mitigation processes that specifically address financial transparency and accountability in recipient countries.[154]

Direct Legal Measures

Involve the International Court of Justice

The International Court of Justice (ICJ) is the UN's principal judicial organ, established under Article 92 of the UN Charter.[155] The Court itself, which is composed of 15 judges, has a twofold role: first, to settle legal disputes between states in accordance with international law, and second, to give advisory opinions on legal matters referred to it by duly authorised UN organs and specialised agencies.[156] The ICJ's greatest strength comes from its important advisory (and norm-setting) function, which has proven important in clarifying issues in atrocity situations, most significantly, the state's obligations to prevent and punish genocide in the case, Bosnia and Herzegovina v Serbia and Montenegro (2007).[157] This case showed that there was indeed a duty of sovereign states to prevent and punish genocide to the extent of their capability.[158] However, the ICJ is limited in that it can only rule on state-to-state issues (not try individuals); and states are only bound by its decisions to the extent that they voluntarily submit to its jurisdiction.[159]

Through its advisory opinions, the ICJ can play a role in encouraging states to strengthen climate action policies, and thereby indirectly protect vulnerable populations from climate-related harm. To date, the ICJ has only been used by vulnerable states to pressure big emitting states to reduce their emissions. For example, since 2000, Pacific Island states have requested an advisory opinion from the ICJ on several occasions concerning the duties of states to ensure that GHG emissions from their territory do not harm other states.[160] The central argument used by smaller states is that unbounded emissions and warming is eroding their human rights. By contrast, there has been little discussion on whether, and how, big emitting countries could use the ICJ, and its advisory opinion particularly, to pressure governments ruling over vulnerable populations to fully implement climate action programmes such as adaptation-focused development programmes with transparent funding allocation, monitoring and evaluation processes.[161] The UK and the US could explore options to work through the UN General Assembly, Security Council, Economic and Social Council, or a variety of special agencies (e.g. World Meteorological Organization, Food and Agriculture Organization, or World Health Organisation), to request an ICJ advisory opinion on whether a particular state is failing in its duty to prevent climate-related harm to its population to the extent of its capability.[162]

Threat of International Criminal Prosecution

The threat of international criminal prosecution can be an effective deterrent against atrocities (and wanton environmental destruction) if there is a real prospect that perpetrators will be tried and punished.[163] There are several legal avenues that international actors could explore to threaten or punish potential perpetrators of atrocities and climate-related damage.

Specialist tribunals have been established on several occasions to deal with war crimes committed in specific conflicts, for example in Rwanda, the Former Yugoslavia, and Sierra Leone (see Chapter 2). The trouble with these tribunals is that by definition they have no direct application outside their country of origin, and the large expense to establish each tribunal tends to deter replication.[164] Specialist tribunals that deal with a range of legal matters in specific contexts beyond atrocities could be a potential forum to address climate-related concerns. For example, the International Tribunal for the Law of the Sea could potentially hear cases on climate-related damage to the marine environment in breach of the 1995 Fish Stocks Agreement;[165] and/or it could address questions arising under the rubric of general international law, for instance: What is the responsibility of a party to the 1982 Convention on the Law of the Sea to prevent sea level rise?[166]

The International Criminal Court (ICC) can investigate and prosecute individuals (unlike the ICJ) for four international crimes: crimes against humanity, war crimes, genocide, and the crime of aggression.[167] Critics of the ICC argue that it is overly Western influenced and African focused, and more than this, in its 20 years in operation, with a total budget of nearly US$2 billion, it has made only ten convictions and four acquittals.[168] This low prosecution rate is largely due to the Court's limited reach, which in the absence of the Security Council referring a particular atrocity situation to the ICC such as in the Darfur case, only extends to the 123 parties to the Rome Statute.[169] Those not party to the Statute include major powers the US, Russia, and China, among other highly relevant countries such as Ethiopia, Israel, Iraq, Libya, North Korea, Sudan, Syria, and Yemen.[170]

The deliberate destruction of the Amazon rainforest in 2020 prompted a renewed push to formally extend the ICC's four crimes to include the crime of 'ecocide',[171] as well as raised the prospect of the ICC investigating political leaders implicated in committing one of the four crimes during an ecocide situation – for example, indigenous leaders allege that under President of Brazil Jair Bolsonaro's watch 'crimes against humanity' were committed during the Amazon wildfires.[172] President of France, Emanuel Macron, has conveyed support for establishing ecocide as the fifth ICC crime, and to add credibility to this international position, in April 2021, the French National Assembly (the lower house of the parliament) passed a bill to make 'ecocide' a national offence carrying jail terms of up to ten years.[173] Western countries beyond France could explore the prospects and limits of establishing ecocide as a national and/or international crime, which in itself may help dissuade recalcitrant governments from intentionally causing large-scale environmental damage, including to the atmosphere.

Universal jurisdiction is another legal avenue to explore. This involves the international legal principle that an individual accused of an atrocity crime can be investigated and prosecuted before a national domestic court irrespective of their nationality or where the crime was committed.[174] This principle was applied in Germany in November 2021 with a former Iraqi member of the Islamic State found guilty of genocide against the Yazidi minority of northern Iraq and sentenced to life in prison.[175] The notion of a 'climate change jurisdiction', which explicitly draws on elements of universal jurisdiction (e.g. the nexus between the 'graveness' and 'universality' of a particular problem), has attracted some scholarly attention but holds limited immediate practical utility.[176]

Other legal options to punish (and therefore deter) alleged perpetrators include: supporting the creation of new international courts that focus on climate-related issues such as an International Environment Court;[177] or working through existing international bodies that can rule on national climate policies that implicate trade law such as the World Trade Organization.[178] Western governments can also support the establishment of domestic environmental courts and tribunals in target countries to adjudicate on climate concerns.[179]

Security Sector Strategies

Structural and Longer-Term Security Measures

Support Security Sector Reform

Security Sector Reform (SSR) involves transforming the armed forces, police and intelligence services to ensure that they are competent and accountable to and representative of broader society.[180] SSR is vital to enhance governance, promote stability, and increase public trust in the state.[181] Ill-disciplined, poorly structured, and otherwise ineffective security forces can inflame tensions and create environments where conflict and atrocities prosper.[182] SSR programmes can be undertaken bilaterally or multilaterally and can involve the provision of equipment and training, institutional reform (e.g. redistributing or consolidating personnel and salaries), as well as strategic planning, and building accountability and oversight mechanisms.[183]

Climate-informed SSR could encompass the implementation of programmes in a target country that directly impacts on the security sector actors themselves as well as indirectly on the political and economic context in which they operate. Direct measures could include establishing climate action issues as a priority in a variety of traditional SSR activities. For example, integrating climate impacts into SSR conflict risk assessment processes,[184] as well as mediation, advisory, and technical services.[185] SSR's training and equipping component could benefit from a climate rendering as well, for example, by ensuring police and military officials have an enhanced understanding about the country's environmental laws and regulations and have access to cutting-edge technology to enforce those rules. As an example, to enable governments in central Africa – home to the second largest carbon sink in the world behind the Amazon – to better enforce industrial logging regulations, USAid has provided satellite remote sensing technology and support for on-ground forest patrols.[186] Finally, SSR and Disaster Risk Reduction programmes can and should be linked because the security sector plays a crucial role in managing and implementing disaster risk reduction and preparedness frameworks. Germany has recently recognised the importance of 'linking DRR to SSR, conceptually and programmatically'.[187]

Measures that could have an indirect effect include integrating SSR and human security programmes.[188] Both concepts developed in the post-Cold War period as a way to reduce conflict, facilitate development and protect individuals rather than nation-states.[189] One of the major criticisms of SSR programmes is that they too narrowly focus on operational fixes (training for example) without addressing the need for institutional and political reform.[190] Human security programmes address these broader reform issues by encouraging human rights, good governance, and better access to healthcare, education, and a clean environment. Conversely, as discussed, the human security agenda could benefit from the field-relevant operational focus inherent in SSR. The US Global Fragility Act of 2019, and subsequent Prologue published by the State Department in 2022, identifies SSR as a priority for US external engagement going forward.[191] Integrating climate issues into this feature of the Act and Prologue would be consistent with the Biden Administration's 'whole of government' approach to US domestic and international climate action.

Transition from Military to Civilian-Controlled Government

Encouraging the transition from a military to civilian-controlled government can deliver improved environmental performance. While non-democratic and democratic governments alike have dismal records on climate and emissions as well as broader environmental protection, countries with military dictatorships tend to have particularly bad environmental records that have improved with a transition to a civilian government. For example, the Brazilian military that was in direct control of the country's federal government between 1964 and 1985 played a prominent role in setting in motion the events that caused the deforestation of vast areas of the Amazon.[192] The so-called slash-and-burn policies of that era, at their peak, led to more than 31,000 km^2 of forest – an area the size of Belgium – being destroyed each year. The new political environment under the New Republic post-1985 allowed activists and politicians to mount a vocal campaign condemning the environmental and social consequences of the regime's approach to the Amazon. A series of reforms in the 2000s caused annual average deforestation to drop to 6400 km^2 per year in 2011.[193] President Bolsonaro's recent, short-lived, unconventional approach to deforestation, as mentioned, seems to contradict this contention. However, Brazil's incoming Lula Administration appears to be reserving these policies, making Bolsonaro's policies the exception to the rule. Regardless, overall, it is likely that as the climate crisis intensifies this century it will become increasingly harder for democratically elected leaders to pursue environmentally damaging policies. This was recently evidenced in the dramatic swing away from a climate-blind Australian Liberal Party towards climate-focused Independent candidates in the Australian general election held in May 2022.[194]

Any successful transition from military to civilian government for environmental or atrocity prevention gains must include constitutional and political change as well as establishing transparent revenue collection and distribution processes.[195] These processes offer multiple entry points for Western governments to include climate action issues, for instance, constitutional provisions can include the right to a clean natural environment as discussed earlier.

Support Confidence-Building Measures

A significant military contribution to conflict prevention, particularly of a cross-border character, can be made by embracing confidence-building measures.[196] Two such measures include reciprocal information exchanges (e.g. formal exchanges about capabilities – troops and other assets, and/or exchanges between agencies and ministers including the establishment of an emergency hotline), and military-to-military programmes (e.g. joint training exercises).[197] The aim of these activities is to reduce mistrust between parties.

Establishing confidence-building measures between states can help protect vulnerable human populations from climate-related disasters. Broad measures could include routine cross-border agency-to-agency information exchanges concerning short-and-long term extreme weather events and forecasts such as severe storms and extreme heat days, combined with direct lines of communication to senior government decision-makers for emergency use. Western countries could facilitate deeper cross-border communication between climate-exposed states by supporting and transferring Disaster Risk Reduction (DRR) activities and technologies such as disaster prevention education, hazard mapping, high-tech meteorological radar and early warning systems, enhanced community-level weather warning communications and evacuation drills, multi-purpose shelters, financial and technical assistance in emergency relief, and post-event psychological care.[198] Countries could also engage in bilateral and regional emergency service and rapid response training, the sharing of capability (personnel and equipment), and response planning. For example, US, Canada and Australia conduct annual joint firefighting training exercises and redistribute personnel and heavy equipment such as aircraft where it is most needed during fire season.[199]

To help develop climate-informed confidence-building measures between climate-exposed states, the US, EU, and UK could consider working through the following entities: their own government agencies (e.g. in the UK, the UK MET Office and UKAid, each of which has a strong DRR outreach programme);[200] established regional bodies (e.g. the South Asian Association for Regional Cooperation – Afghanistan, Bangladesh, Bhutan, India, the Maldives, Nepal, Pakistan, and Sri Lanka – which has attempted to improve institutional mechanisms for different facets of DRR preparedness and response);[201] task-specific groupings (e.g. the Coalition for Disaster Resilient Infrastructure, which brings together 31 countries including the US and UK (and the EU), several UN agencies, and financial institutions);[202] or international organisations (e.g. the UN Office of DRR, which is the UN's focal point for disaster risk reduction activities).[203]

Adherence to Regimes and Treaties

Another significant structural or longer-term contribution to conflict prevention can be made by encouraging states to adhere to the full range of arms control, disarmament, and non-proliferation regimes and treaties, with a special focus on supply-side control regimes over light and small-scale arms, given their highly destructive role in internal conflicts.[204] Other notable treaties highly relevant to internal conflicts include the 1993 Chemical Weapons Convention, 1997 Mine Ban Treaty, and 2008 Convention on Cluster Munitions – all of which prohibit the production, stockpiling, and use of the respective lethal technology.[205] Regimes and treaties help guide what is considered appropriate and inappropriate behaviour in international relations. Both involve states coming together in a formalised way to advance a particular issue of international concern; however, unlike regimes, a treaty has a legal quality to it.

With these options in mind, Western governments could help protect vulnerable populations from climate-related harm in the following ways. First, by recommitting themselves to existing emission reduction treaties such as the Paris Agreement and its temperature goal of limiting warming to under 2°C. Recent analysis suggests that holding temperature rise to 1.5°C is improbable, but 2°C remains feasible.[206] Second, by encouraging countries to adhere to the regimes and treaties that seek to limit damage to natural systems. There are more than 2000 multilateral and bilateral environmental treaties and amendments covering atmospheric pollution, biodiversity, chemicals, waste, land, and water issues, that could be relevant.[207] Third, by working with fragile countries in climate hotspots to help resolve resource-sharing issues and secure appropriate frameworks to manage those resources. Agreements and treaties on shared water resources offer especially important learnings such as to avoid politicising natural resources that are essential for human rights and national security.[208] Fourth, by developing new regional regimes and treaties to deal with emissions at their source, for example, by establishing a non-fossil fuel proliferation regime.[209] Fifth, by leading by example on existing and new climate-related certifications and standards such as those related to forest protection and hydrogen development.[210] On the latter, leadership requires supporting certification regimes that identify and promote 'green' hydrogen produced with renewable energy; not certification regimes that blend renewable hydrogen with 'blue' (gas) or brown (coal) hydrogen to produce 'environmentally friendly' hydrogen – this approach simple extends the life of fossil fuels in the global energy system.

Direct Security Measures

Threaten Arms Embargoes

Arms embargoes are a type of sanction that can be used to coerce states and non-state actors to improve their behaviour in the interests of peace and security.[211] They involve prohibiting the export or supply of conventional military equipment and/or 'dual-use goods' as well as any associated financing and technical assistance. Arms embargoes can be applied by individual states, regional organisations or the international community – under Chapter VII authority of the UN Security Council – as a preventive and reactive response to a human protection catastrophe.[212] Western governments regularly threaten and/or apply arms embargoes on various entities.

'Dual-use goods' (specific items that can be used to achieve military and peaceful goals)[213] are critical to successfully implement a climate action agenda. Items that fall within the dual-use goods category can include a variety of electronic and computer components and systems, global positioning and navigation systems, aerospace technology, thermal, sensor, and laser equipment, and chemical, biological, and nuclear technologies and materials.[214] The climate-related tasks that these dual-use goods perform include, for example, tracking weather systems and providing early warnings should a severe storm approach, search and rescue functions, and firefighting. In addition, access to innovative technologies such as machinery and chemicals are essential to build competitive low-emission economies. Often restricting the export of dual-use goods serves as an additional measure to complement traditional arms embargoes, and the

US, UK, and EU provide exceptions for dual-use goods that serve humanitarian needs. Careful analysis would need to be undertaken to avoid doing more harm than good to the population at large should this option be pursued.

Withdraw Military Cooperation

Intertwined with arms embargoes is the option to withdraw from military cooperation programmes with a target country. This option can elicit behavioural change in the interests of human protection, especially if the country in question is highly dependent on external defence programmes for its security. The US Department of Defense (DoD) conduct 'security cooperation' programmes in more than 130 countries.[215] These programmes range from the expensive and visible sale of military equipment and training, and conducting joint exercises, to low-key bilateral talks, workshops, personnel exchanges, and professional military education.[216] DoD security cooperation with developed countries tends to focus on enhancing interoperability with US systems, whereas with developing and fragile countries, security cooperation generally focuses on building security force capacity to improve its day-to-day operational capabilities.[217]

Foreign governments can purchase US defence items though the DoD's Foreign Military Sales programme.[218] Items range from high-end aircraft and missile systems to light arms and tactical clothing. However, items for sale also include those essential to protect populations from the impacts of climate change, for example: radar and military electronics; planes, helicopters, and refuelling tankers; patrol boats; firefighting equipment including vehicles; rescue and safety equipment; water purification equipment; as well as spare parts, and repair and return programmes; and accompanying facility construction and logistics, training support and exercises, technical assistance, and joint research.[219] As a practical example, US security cooperation with Bangladesh includes funding for marine assets, joint training exercises and exchanges, and support for training, equipment, and facilities upgrades, in part to enhance Bangladesh's 'humanitarian assistance and disaster response capabilities'.[220] As with embargoes on dual-use items, withdrawing lifesaving equipment and staff risks doing more harm than good in some instances.

Nonterritorial Show of Force

A nonterritorial show of force involves deploying military resources without an actual intervention occurring on the territory of the targeted state – for example, positioning warships off the coast of a target state – hence its colloquial name: 'gunboat diplomacy'.[221] This option avoids the need to secure the consent of the target state, brings media attention to the issues at hand, and adds weight to any diplomatic initiatives that should always accompany such an option. It can also serve as an instrument to monitor and enforce sanctions and arms embargoes, help deter trouble (e.g. discourage spoilers during peace negotiations), and provide a rapid response capability should trouble arise. However, it risks being interpreted as deeply threatening and consequently may provoke an unanticipated reaction by the target state and/or its allies.

In the aftermath of a high-impact climate event such as a major cyclone, a variety of military assets can and should be immediately deployed in order to perform essential lifesaving tasks such as search and rescue and the distribution of emergency medical and food supplies, temporary accommodation, and disaster responders. Countries that are suspicious of Western intentions may strongly resist emergency assistance, particularly if delivered by foreign militaries, as was the case in the aftermath of Cyclone Nargis that made landfall in May 2008 in Myanmar.[222] However, suspicions and paranoia by those in the government, should not deter Western countries from pursuing a policy that marshals lifesaving military assets such as helicopters, planes, warships, and personnel, in border countries and off the coast of a specific country in order to save lives. Accompanying diplomatic engagement with members of the target government is essential.

Preventive Deployment

Preventive deployment involves positioning troops (and/or military observers, civilian police and assets) either within or across state borders where there is an emerging threat of conflict, with the consent of the government or governments involved, for the primary purpose of deterring the escalation of that situation into armed conflict.[223] This type of deployment draws international attention to the issues and actors involved, and at least implies a willingness to take further action should the situation deteriorate.[224] Mission tasks can include monitoring and reporting on the situation as well as setting the conditions for diplomatic measures to commence such as mediation and negotiation.

Preventive deployment could be considered in circumstances where a climate-related event is eroding, or had eroded, a subnational balance-of-power that credibly may result in large-scale harm being committed against a minority population. For example, where capacity is low, a central government could request a preventive deployment approved by the Security Council to deter an armed militia group from forcibly removing inhabitants from lands and territories in the aftermath of a major cyclone or flood – the event itself viewed by the militia group as a window of opportunity to advance their strategic interests and to settle old scores. A preventive deployment could also be considered to deter climate-related cross-border disputes, such as those involving access to shared water resources, from escalating into violent conflict. In these situations, deployed personnel could serve as a critical early-warning alarm for the Security Council, help to strengthen mutual dialogue among political parties, and assist in monitoring human rights as well as inter-ethnic relations in the country.[225]

Notes

1 Article 33, United Nations Charter, Chapter VI: Pacific Settlement of Disputes.
2 Boutros Bourtos-Ghali, *An Agenda for Peace: Preventive Diplomacy, Peacemaking, and Peace-keeping*. Report of the Secretary-General (United Nations, 1992), paragraphs 15 and 23.
3 Kofi Annan, 'Prevention of Armed Conflict', Report of the Secretary-General, A/55/985 and S/2001/574, *General Assembly*, p 36.
4 *The Responsibility to Protect*, ICISS, p xi and 19–23.
5 'World Summit Outcome', A/RES/60/1, 24 October 2005.
6 'United Nations Convention on Climate Change', *UNFCCC*, 1992, p 4.
7 'What Is the United Nations Framework Convention on Climate Change?', *UNFCCC* (accessed 2020).
8 Kyoto Protocol, *UNFCCC*, 1998.
9 Copenhagen Accord, *UNFCCC*, 2009.
10 Paris Agreement, *UNFCCC*, 2015.
11 Glasgow Climate Pact, *UNFCCC*, 2022.
12 Boutros-Ghali, *An Agenda for Peace*, Item 26 and 27. See also 'Preventing Violence, War and State Collapse: The Future of Conflict Early Warning and Response', *OECD*, 2009.
13 Mads Knudsen, 'Integrating Climate Change and Conflict Early Warning: A Project Proposal', Issue Brief 17, UN Development Program, 2020, p 8. See also, Ulf Engel, 'The African Union's Conflict Early Warning System Is No More. What Now?', *The Conversation*, 24 May 2022.
14 *The Responsibility to Protect*, ICISS, p 21.
15 Ibid., p 22.
16 'World Summit Outcome', p 30.
17 'Framework of Analysis for Atrocity Crimes: A Tool for Prevention', *United Nations*, 2014, p 5.
18 Ibid.
19 Ibid.
20 Ibid., p 4.
21 'Climate Change and Atrocity Crimes: The Challenge in the Pacific', *Asia-Pacific Centre for Responsibility to Protect*, November 2020.
22 Ibid., p 4.
23 Ibid., p 18.
24 Simon Adams, *Mass Atrocities, the Responsibility to Protect and the Future of Human Rights: 'If Not Now, When?'* (Oxon: Routledge, 2021), p 87.
25 'Climate Action. Early Warning Systems', *United Nations* (accessed 2022).
26 'Multi-hazard Early Warning Service (MHEWS) Tanzania', *Met Office* (accessed 2022).
27 Jessica Caus and Adam Day, 'A Storm on the Horizon: How to Translate Climate Security Early Warning into Action', *PeaceLab*, 23 February 2021.
28 'Water Peace and Security' (accessed 2022). See also reporting on the WPS at 'Water Wars: Early Warning Tool Uses Climate Data to Predict Conflict Hotspots', *The Guardian*, 08 January 2020.
29 'ClimSec Sahel Early Warning System on Human Security in the Sahel Region: Considering Climate Impacts on Food Security and Farmer Herder Conflicts', *Potsdam Institute for Climate Impact Research* (accessed 2022).

'Weathering Risk: A Climate and Security Risk and Foresight Assessment: Project Overview', *Adelphi* (assessed 2022).
30 'Climate Action. Early Warning Systems' (accessed 2022).
31 'WMO Congress Boosts Early Warning and Humanitarian Action', *World Meteorological Organisation*, 14 June 2019.
32 'Risk-Informed Early Action Partnership' (accessed 2022).
33 'Weathering Risk' (assessed 2022).
34 Jessica Caus and Adam Day, 'A Storm on the Horizon: How to Translate Climate Security Early Warning into Action', *PeaceLab*, 23 February 2021.
35 See for example, Gareth Evans, *The Responsibility to Protect: Ending Mass Atrocity Crimes Once and For All* (Washington, DC: Brookings Institution Press, 2008), p 84.
36 Ibid.
37 Ibid.
38 'Climate Security Mechanism: Toolbox', Briefing Note, *Department of Political and Peacekeeping Affairs* with the *UN Development Programme* and *UN Environment Programme*, 2019, p 9.
39 See Part 2 'Operationalizing the Responsibility to Protect' in Evans, *The Responsibility to Protect*, p 78–174. Some policy options are not covered such as the notion of 'legitimacy' in R2React and the full range of legal options in R2Rebuild.
40 Evans, *The Responsibility to Protect* (Brookings, 2008), p 88.
41 Ibid.
42 Ibid.
43 Harriet Bulkeley and Peter Newell, *Climate Change Governance* (London: Routledge, 2010).
44 See for example, 'Council Conclusions on Climate Diplomacy – Council Conclusions', 6094/18, *European Union*, 26 February 2018; 'Biden Introduces His Climate Team', *The New York Times*, 22 December 2022.
45 See for example, Johara Bellali, Lisa Strauch, Francis Oremo, and Benson Ochieng, 'Multi-Level Climate Governance in Kenya: Activating Mechanisms for Climate Action', *Adelphi* and *The Institute of Law and Environmental Governance,* October 2018.
46 The US's Foreign Assistance programme appropriated USD$746 million in 2019 to support 'good governance' worldwide, which does not appear to include a climate governance category. In the period 2019–2020, the UK Foreign Commonwealth and Development Office ('Development Tracker') shows that the UK allocated 1033 million pounds to 'Government and Civil Society', similarly however, climate governance support does not feature as a specific category.
47 Climate Action Tracker provides a methodology to assess climate governance, see 'Climate Governance: Assessing Governments Ability and Readiness to Transform their Societies into Zero Emissions Societies', Methodology Note, Climate Governance Series, *Climate Action Tracker*, December 2020.
48 Evans, *The Responsibility to Protect* (Brookings, 2008), p 89.
49 Ibid.
50 Ibid.
51 'The EU Emissions Trading System (EU ETS)', Fact Sheet, *European Commission* (accessed 2022).
52 'Climate Change: What Is the EU Doing?', *European Council* (accessed 2022).
53 'Carbon Pricing', *The World Bank* (accessed 2022).

54 NATO Secretary General Jens Stoltenberg, 'NATO Must Combat Climate Change', Opinion Editorial, *Die Welt*, 29 September 2020.
55 Evans, *The Responsibility to Protect* (Brookings, 2008), p 89.
56 Ibid.
57 Ban Ki-moon, *Preventive Diplomacy: Delivering Results*, Report of the Secretary-General, 2011, p 9. Asif Khan, Chief of Mediation Support and Gender, Peace and Security within the Policy and Mediation Division in the DPPA, explains that the Good Offices can mean

> any third-party assistance given to conflict parties to help find a solution to their problems. It can involve advising parties to a conflict. Or advising governments. It can mean carrying messages between opposing sides. And it can mean trying to build contact and trust between groups. And sometimes it can mean providing different expertise on difference aspects of a peace process.
>
> ('What are the UN's "Good Offices" and Successful Mediation?', *YouTube*, 19 February 2020)

58 In addition to the more recent list of speeches identified in chapter 3, see also, the UN Secretary-General António Guterres: 'Remarks on Climate Change', 10 September 2018; 'Climate Change Is Biggest Threat to Global Economy', 25 January 2019; 'Message on the International Day of Peace', 21 September 2019; 'Remarks at Climate Action Summit', 23 September 2019; 'Remarks at pre-COP25 Press Conference', 1 December 2019; 'Remarks to United Nations Biodiversity Summit', 30 September 2020; and 'The State of the Planet', 2 December 2020.
59 See for example, DPPA Under-Secretary-General Rosemary DiCarlo, 'Climate Change Multiplying Factors that Lead to Insecurity for Millions, Rosemary DiCarlo tells "Arria Formula" Meeting', United Nations, 22 April 2020.
60 'Climate Security Mechanism Toolbox – Conceptual Approach', *DPPA* (accessed 2022). See also, 'Strategic Plan: United Nations Department of Political and Peacebuilding Affairs; 2020–2022', *DPPA*, p 29.
61 See separate documents: 'Climate Security Mechanism: Toolbox', 'Briefing Note' and 'Conceptual Approach'. See also, 'Annual Report: Multi-Year Appeal', *DPPA*, 2019, p 28.
62 'Annual Report: Multi-Year Appeal', *DPPA*, 2019, p 28. Political Affairs Officer DPPA, Thomas Ritzer, 'What Is the Link Between Climate Change and Conflict?', *YouTube*, 15 April 2020.
63 'United Nations Peacemaker' (accessed 2022).
64 See United Nations: Guidance for Effective Mediation (New York: United Nations, 2012).
65 Other past examples include, From Conflict to Peacebuilding: The Role of Natural Resources and the Environment (2009); Protecting the Environment During Armed Conflict: An Inventory and Analysis of International Law (2009); Greening the Blue Helmets: Environment and Natural Resources in UN Peacekeeping Operations (2012); Women and Natural Resources: Unlocking the Peacebuilding Potential (2013); The Role of Natural Resources in Disarmament, Demobilization and Reintegration (2013).
66 The Implications of Climate Change for Mediation and Peace Processes: DPPA Practice Note DPPA, September 2022.
67 'Concept on EU Peace Mediation', 13951/20, *Council of the European Union*, 11 December 2020, p 8; 'Peace Mediation Guidelines', *European External Action Service*, December 2022, p 23–24.

68 'United Nations Climate Security Mechanism', Progress Report, May 2022, p 14.
69 Evans, *The Responsibility to Protect* (Brookings, 2008), p 90.
70 Ibid.
71 See for example, 'UK Sanctions Regimes: Information on UK Sanctions Regimes Currently in Force', *UK Government* (accessed 2022).
72 'Global Human Rights Sanctions', *UK Government* (accessed 2022).
73 'Guidance: UK Sanctions Regime', *UK Government* (accessed 2022). See also, 'Trade Sanctions, Arms Embargoes, and other Trade Restrictions', *UK Government* (accessed 2022).
74 See for example, Michelle Bachelet, United Nations High Commissioner for Human Rights, Opening Statement to the 42nd session of the Human Rights Council, 9 September 2019.
75 Ibid.
76 'Global Magnitsky Act', *US State Department* (accessed 2022).
77 Richard Youngs, 'The New EU Global Human Rights Sanctions Regime: Breakthrough or Distraction?', *Carnegie Europe*, 14 December 2020.
78 See Evans, *The Responsibility to Protect* (Brookings, 2008), p 91.
79 Ibid., p 91–92.
80 See for example, Foreword to 'Turn Down the Heat: Why a 4C Warmer World Must be Avoided', A Report for the World Bank by the Potsdam Institute for Climate Impact Research and Climate Analytics, 2012; 'World Bank Group President Jim Yong Kim's Statement at the Spring Meetings 2013 Opening Press Conference', 18 April 2013; 'Statement by World Bank Group President Jim Yong Kim on the appointment of Michael R. Bloomberg as the Special Envoy for Cities and Climate Change to the United Nations', 31 January 2014; 'Remarks by World Bank Group President Jim Yong Kim at the Announcement of the Transformative Carbon Asset Facility', 30 November 2015. And more recently, 'Climate Change – Overview', World Bank, 2020.
81 'Sustainable Development, Poverty Eradication and Reducing Inequalities' (Chapter 5), in *Global Warming of 1.5°C. An IPCC Special Report on the Impacts of Global Warming of 1.5°C*, IPCC, October 2018.
82 'Climate Change Threatens Progress Across Sustainable Development, Warns New UN Report', Press Release, UN SDG, 9 July 2019.
83 See for example, ' "Green Steel": The Race to Clean Up One of the World's Dirtiest Industries', *Financial Times*, 15 February 2021; 'Sustainable Cement: The Simple Switch that Could Massively Cut Global Carbon Emissions', *World Economic Forum*, 27 August 2022.
84 See for example, 'Petersburg Climate Dialogue: Green Recovery Will Lead to a More Resilient Future', *UNFCCC*, 28 April 2020; 'UK Support Grows for a Green Covid-19 Exit', *Climate News Network*, 28 May 2020; 'Make Covid-19 Recovery Green, say Business Leaders', *BBC*, 1 June 2020; 'Building Back Better: A Sustainable, Resilient Recovery after COVID-19', *OECD*, 5 June 2020; 'Greening the Recovery', *IMF*, 2020.
85 Jocelyn Timperley, 'The Broken $100-Billion Promise of Climate Finance — and How to Fix It', *Nature*, 20 October 2021.
86 The Debate about Climate Compensation Has Been Simmering Away for Decades see for example, Jeffrey Sachs, 'Climate Change and War', *Project Syndicate*, 1 March 2005.
87 Matt McDonald, 'COP27: One Big Breakthrough but Ultimately an Inadequate Response to the Climate Crisis, *The Conversation*, 20 November 2022.

88 Adil Najam, 'COP27's 'Loss and Damage' Fund for Developing Countries Could be a Breakthrough – Or Another Empty Climate Promise', *The Conversation*, 22 November 2022.
89 Evans, *The Responsibility to Protect* (Brookings, 2008), p 92.
90 Ibid., p. 92–93.
91 Ibid., p. 93.
92 'What Ties Us Together: Impact Report, 2019–2020', *Search for Common Ground*, 2020, p 3.
93 Karen E McNamara et al., 'An Assessment of Community-Based Adaptation Initiatives in The Pacific Islands', *Nature Climate Change*, Vol 10, June 2020, p 628.
94 McNamara et al., 'An Assessment of Community-Based Adaptation'.
95 'The GEF Small Grants Programme: Community Based Adaptation', *GEF* and *UNDP* (accessed 2022).
96 'Community-Based Adaptation in Practice: A Global Overview of CARE International's Practice of Community-Based Adaptation (CBA) to Climate Change', *CARE International*, 2014.
97 See for example, 'Pacific Islands Coastal Community Adaptation Project, 2012–2019', Final Report, USAid.
98 Evans, *The Responsibility to Protect* (Brookings, 2008), p 93.
99 Ibid.
100 Richard B. Stewart, Benedict Kingsbury, and Bryce Rudyk, 'Climate Finance: Key Concepts and Ways Forward', Harvard Project on International Climate Agreements, 2017.
101 'Global Climate Finance: An Updated View 2018', *Climate Policy Initiative*, November 2018. Also see Sophie Yeo, 'Where Climate Cash Is Flowing and Why It's Not Enough', *Nature*, 17 September 2019.
102 'USAID Climate Change and Development Strategy, 2012–2018', Report, *USAid*, 2012. 'UK to Put Climate Crisis and Environment at the Heart of Overseas Aid', *The Guardian*, 3 July 2019; 'Aid Funding Must Recognise Climate Change Emergency, say MPs', *The Guardian*, 8 May 2019; *UK Aid for Combatting Climate Change*, Report, House of Commons, International Development Committee, 8 May 2019. 'UK Aid for Combating Climate Change: Government Response to the Committee's Eleventh Report', UK Parliament, 18 July 2019.
103 Barack Obama, *A Promised Land* (New York: Crown Publishing Group, 2020), p 514.
104 'Reducing UK Emissions: Progress Report to Parliament', UK Committee on Climate Change, p 102. See also, 'UK Makes Assurances on Climate Finance Amid Concerns Over Aid Restructuring', *Climate Home News*, 26 July 2020.
105 'UK Aid to Double Efforts to Tackle Climate Change', Press Release, UK Government, 23 September 2019.
106 Evans, *The Responsibility to Protect* (Brookings, 2008), p 94.
107 'Trade Sanctions, Arms Embargoes, and Other Trade Restrictions', UK Government (accessed 2022).
108 Article 4.5 in 'United Nations Framework Convention on Climate Change', United Nations, 1992.
109 See for example, John H. Barton, 'Mitigating Climate Change Through Technology Transfer: Addressing the Needs of Developing Countries', *Chatham House*, October 2008; 'Climate Technology Transfer', *European Commission*, 2020; Hendrick Larsen, 'Technology Transfer: How the UNFCCC Helps Put the

Right Technologies into the Right Hands', *Grantham Institute*, 12 December 2016; 'Technology Mechanism', *UNFCCC*, 2022.

110 'UK to Stop Funding Overseas Fossil Fuel Projects in Climate Summit Vow', *The Sydney Morning Herald*, 12 December 2020. See also, 'Carbon Emissions: Scale of UK Fossil Fuel Support "Staggering"', *BBC*, 23 January 2020.

111 'Rich Nations Vowed Billions for Climate Change. Poor Countries Are Waiting', *The New York Times*, 9 September 2018. Under the Paris Agreement developing countries by and large have a two emission reduction targets, the stronger of which is contingent on external support. For example, Chile has pledged to 'unconditionally' reduce its GHG emissions by 30 per cent in 2030, but with external financing and technical support, it will reduce emissions by 45 per cent.

112 'How Trump Could Slow Climate Change Projects Around the World', *Time*, 13 June 2017.

113 Evans, *The Responsibility to Protect* (Brookings, 2008), p 94.

114 David Cortright (ed.), *The Price of Peace: Incentives and International Conflict Prevention* (Lanham: Rowman and Littlefield, 1997), p 10.

115 Paulina Matera, 'Economic Incentives as Tools of Foreign Policy: Polish Engagement in Ukraine', *European Review*, Vol 30, No 1, p 12–13. See also, William J. Long, 'Trade and Technology Incentives and Bilateral Cooperation', *International Studies Quarterly*, Vol 40, 1996, p 77–106.

116 This list of options, excluding the climate discussion, can be found in Matera, 'Economic Incentives as Tools of Foreign Policy', p 4–11.

117 Evans, *The Responsibility to Protect* (Brookings, 2008), p 95.

118 Ibid.

119 'Constitution Transformation Network', Melbourne Law School, The University of Melbourne (accessed 2022).

120 Evans, *The Responsibility to Protect* (Brookings, 2008), p 95.

121 *Constitution of the Republic of South Africa*, 1996, Section 24.

122 'Multi-Level Climate Governance in South Africa: Catalysing Finance for Local Climate Action', One-World Sustainable Investments, *Sustainable Energy Africa* and *Adelphi*, January 2019, p 26.

123 *The Constitution of Kenya*, 2010, Section 42. For a comprehensive assessment of Kenya's constitution making processes see work undertaken by Christina Murray – a constitutional scholar, senior mediation advisor for the UN, and member of the Kenyan Constitution Committee of Experts.

124 Johara Bellali, Lisa Strauch, Francis Oremo, and Benson Ochieng, 'Multi-level Climate Governance in Kenya: Activating Mechanisms for Climate Action', *Adelphi* and *the Institute of Law and Environmental Governance*, October 2018, p 9–10 and 30.

125 'Guidance Note of the Secretary-General on United Nations Constitutional Assistance', *United Nations*, September 2020, p 1.

126 Ibid.

127 'The Implications of Climate Change for Mediation and Peace Processes: DPPA Practice Note', *DPPA*, September 2022, p 17.

128 'Standby Team of Senior Mediation Advisers', *UN Peacemaker* (accessed 2022).

129 Evans, *The Responsibility to Protect* (Brookings, 2008), p 96.

130 Ibid.

131 UN Secretary-General António Guterres, 'Remarks to the UN Human Rights Council: "The Highest Aspiration: A Call to Action for Human Rights"', 24 February 2020.

132 'OHCHR and Climate Change', *UN Office of the High Commissioner for Human Rights* (accessed 2022).
133 Joel E. Correia, 'The UN Declared a Universal Human Right to a Healthy, Sustainable Environment – Here's Where Resolutions Like This Can Lead', *The Conversation*, 5 August 2022.
134 This list has been adapted to reflect a climate aspect from the list of activities in Evans, *The Responsibility to Protect* (Brookings, 2008), p 97.
135 See for an overview 'Amnesty International: Climate Change' (accessed 2022); and 'Human Rights Watch: Climate Change' (accessed 2022).
136 Mary Robinson, 'Rights for Action: Putting People at the Centre of Action on Climate Change', *Mary Robinson Foundation for Climate Justice*, 2015, p 2. See also Mary Robinson, 'Why Climate Change Is a Threat to Human Rights', TED Talks.
137 Evans, *The Responsibility to Protect* (Brookings, 2008), p 97.
138 Ibid., p 98.
139 Ibid.
140 Cross check 'Populations at Risk', *Global Centre for the Responsibility to Protect* (accessed 2022) and 'Climate Change Laws of the World', *London School of Economics* (accessed 2022). Findings: Burkina Faso, China, Democratic Republic of Congo, Myanmar, Niger, Nigeria, North Korea Yemen have general laws on environmental protection and/or climate change/air pollution; Afghanistan, Burkina Faso, Cameroon, China, Democratic Republic of Congo, Ethiopia, Mali, Myanmar, Nigeria, North Korea, Syria, Yemen have laws governing their electricity sector (including renewable electricity); Central African Republic and Syria have laws on liquid fuels (and by implication, transport); Afghanistan, Central African Republic, China, Democratic Republic of Congo, Ethiopia, Mali, Myanmar, Syria have laws related to land-use and land-use change (forestry and agriculture); and Afghanistan, Burkina Faso, Ethiopia, Mozambique, and Myanmar have laws on climate-related disaster management.
141 'U.S. Climate Change Litigation', *Columbia Law School* and *Arnold & Porter* (accessed 2022). Cross check 'Populations at Risk', *Global Centre for the Responsibility to Protect* (accessed 2022) and Climate Change Laws of the World', *London School of Economics* (accessed 2022).
142 See 'The Biden Plan to Secure Environmental Justice and Equitable Economic Opportunity'.
143 Evans, *The Responsibility to Protect* (Brookings, 2008), p 98.
144 Ibid.
145 Ibid.
146 Ibid.
147 'What We Do', *Transparency International* (accessed 2022).
148 'What Does the Climate Crisis Have to Do with Corruption?', *Transparency International* (accessed 2022). See also, 'International Anti-Corruption Day 2019: Time to Act Against Corruption and the Climate Crisis', *Transparency International*, 5 December 2019.
149 'International Anti-Corruption Day 2019'.
150 'What Does the Climate Crisis Have to Do with Corruption?'.
151 'International Anti-Corruption Day 2019'.
152 'Population at Risk', *Global Centre for R2P* (accessed 2022); 'Corruption Perceptions Index', *Transparency International* (accessed 2022); 'ND-GAIN Country Index', *University of Notre Dame* (accessed 2022).

153 'Contributors', *Climate Funds Update* (accessed 2022).
154 'What Does the Climate Crisis Have to Do with Corruption?'.
155 'United Nations Charter, Chapter XIV: The International Court of Justice, Article 94', *United Nations*, 2022.
156 'Frequently Asked Questions', *International Court of Justice* (accessed 2022).
157 Evans, *The Responsibility to Protect* (Brookings, 2008), p 99.
158 Ibid., p 20.
159 'United Nations Charter, Chapter XIV: The International Court of Justice, Article 94', *United Nations*, 2022.
160 Dan Bodansky, 'The Role of the ICJ in Addressing Climate Change: Some Preliminary Reflections', Public Lecture, *King's College London*, 13 June 2017. 'Students Want International Court of Justice to Rule on Climate Change', *ABC News*, 29 July 2019. For detailed analysis see 'Climate Change and The International Court of Justice', *Yale Centre for Environmental Law and Policy*, 2013.
161 Countries involved in ICJ-climate discussions such as Vanuatu, Tuluva, and Palau, as well as leading legal voices on this matter, do not seem to address this issue, see for example Philippe Sands, 'Climate Change and the Rule of Law: Adjudicating the Future in International Law', *Journal of Environmental Law*, Vol 28, 2016, 19–35; Daniel Bodansky, 'The Role of The International Court of Justice in Addressing Climate Change: Some Preliminary Reflections', *Arizona State Law Journal*, Vol 49, 2017, p 689–712.
162 UNGA has requested an advisory opinion 16 times; WHO and ESC twice each; UNSC once; WMO and FAO have never made a request. 'Organs and Agencies Authorized to Request Advisory Opinions', *International Court of Justice*, 2022.
163 Evans, *The Responsibility to Protect* (Brookings, 2008), p 99.
164 Ibid.
165 Bodansky, 'The Role of The International Court of Justice in Addressing Climate Change', p 700. For more detail see William C.G. Burns, 'Potential Causes of Action for Climate Change Impacts Under the United Nations Fish Stocks Agreement', *Sustainable Development Law and Policy*, Winter, 2007, p 34–38.
166 Sands, 'Climate Change and the Rule of Law', p 25 and 33.
167 'How the Court Works', *International Criminal Court* (accessed 2022).
168 Matt Killingsworth, '20 Years on, The International Criminal Court Is Doing More Good Than Its Critics Claim', *The Conversation*, 12 July 2022.
169 'Darfur, Sudan', *International Criminal Court* (accessed 2022).
170 Claire Klobucista, 'The Role of the International Criminal Court', *Council on Foreign Relations*, 28 March 2022.
171 'International Lawyers Draft Plan to Criminalise Ecosystem Destruction', *The Guardian*, 30 November 2020; 'How 165 Words Could Make Mass Environmental Destruction an International Crime', *NPR*, 27 June 2021. See also, Malayna Raftopoulos and Joanna Morley, 'Ecocide in the Amazon: The Contested Politics of Environmental Rights in Brazil, *The International Journal of Human Rights*, Vol 24, No 10, 2020, p 1616–1641.
172 'Jair Bolsonaro Could Face Charges in The Hague Over Amazon Rainforest', *The Guardian*, 23 January 2021; 'Brazilian Amazon: Crimes Against Nature Are Also Crimes Against Humanity', *The Brussels Times*, 9 November 2022.
173 '"Ecocide" Proposal Aiming to Make Environmental Destruction an International Crime', *ABC News*, 13 February 2021; 'France Drafts 'Ecocide' Bill to Punish Acts of Environmental Damage', *France 24*, 14 April 2021.

174 Evans, *The Responsibility to Protect* (Brookings, 2008), p 100.
175 'Yazidi Genocide: IS Member Found Guilty in German Landmark Trial', *BBC*, 30 November 2021.
176 Natalie L. Dobson, 'Exploring the Crystallization of 'Climate Change Jurisdiction': A Role for Precaution?', *Climate Law*, 8, 2018, p 219. See also, Natalie L. Dobson and Cedric Ryngaert, 'Provocative Climate Protection: EU 'Extraterritorial' Regulation of Maritime Emissions', *International and Comparative Law Quarterly*, Vol 66, April 2017, p 306.
177 Alexander M. Solntsev, 'The International Environmental Court – A Necessary Institution for Sustainable Planetary Governance in the Anthropocene', in Michelle Lim (ed.) *Charting Environmental Law Futures in the Anthropocene* (Bristol: Springer, 2019), p 129–138. See also, Rob White, *Climate Change Criminology* (Bristol University Press, 2018), p 127, 137 and 151; and the 'International Court for the Environment (or "ICE Coalition")' advocacy group (accessed 2022).
178 Bodansky, 'The Role of The International Court of Justice in Addressing Climate Change', p 700.
179 George Pring and Catherine Pring, 'Environmental Courts and Tribunals: A Guide for Policy Makers', Report, *UNEP*, September 2016.
180 This formulation combines Gareth Evans's from his 2008 book on R2P and the multiple other definitions on offer. See for example 'SSR Overview', *Geneva Centre for Security Sector Governance* (accessed 2022); 'Security Sector Reform (SSR)', *United Nations and the Rule of Law* (accessed 2022). UNDPPA, CSM, Toolbox Checklist.
181 Evans, *The Responsibility to Protect* (Brookings, 2008), p 100.
182 Ibid.
183 'Security Sector Reform in DRC: 10 Years of Analysis and Recommendations for Ways Forward', 2018.
184 For example, seek to integrate the questions to consider in the UN's report 'Climate Security Mechanism: Toolbox Checklist' with the questions found in the Appendix of the US Government's report 'Effective Justice & Security Sector Assistance in Conflict-Affected Areas Guidelines for U.S. Government Assistance Planning, Design, And Implementation, 2019–2020', which does not mention climate change.
185 See for example integration opportunities between the 'Security Sector Reform', *UN Peacekeeping*, which does not mention climate change; and the Department of Political and Peacekeeping Affairs 'climate-informed mediation' focus.
186 'Democratic Republic of Congo: The Environment', *USAid* (accessed 2022). See also, '12 Rangers Among 17 Killed in Congo Park Ambush', *The New York Times*, 25 April 2020; 'Gorillas, Guns, and Oil: A Journey Inside Virunga National Park', *The Atlantic*, 25 June 2014; 'US Army Trains Congo Soldiers', *VOA*, 28 October 2010.
187 Viola Csordas and Camilla Arvastson, 'A Comprehensive Understanding of Security Sector Governance Makes for More Effective Reform', *PeaceLab*, 21 April 2020. See also, 'Disaster Risk Reduction, Preparedness and Relief & the Security Sector', *Geneva Centre for Security Sector Governance* (accessed 2022).
188 'SSR in a Nutshell: Manual for Introductory Training on Security Sector Reform', *Geneva Centre for the Democratic Control of Armed Forces*, 2012, p 1–3.
189 Ibid., p 1.

190 'Security Sector Reform in DRC: 10 Years of Analysis and Recommendations for Ways Forward', 2018; Rigobert Minani Bihuzo, S.J., 'Security Sector Reform (SSR) Challenges to Peace in the Democratic Republic of the Congo', African Program Occasional Paper, *Wilson Centre*, July 2020.

191 'Prologue to the United States Strategy to Prevent Conflict and Promote Stability', *State Department* (accessed 2022). See also, Daniel Mahatny, 'U.S. Support to Security Sector Reform in Fragile States', *Center for Strategic and International Studies*, 26 August 2020.

192 Charles Wood and Marianne Schmink, 'The Military and the Environment in the Brazilian Amazon', *Journal of Political and Military Sociology*, Vol 21, Summer 1993, p 81.

193 See Ben L. Parr, Carlos F. Young, and Don Henry, 'Opportunities for a Low Pollution Economic Recovery in Brazil', Briefing Paper, *University of Melbourne*, May 2017.

194 Adam Morton and Katharine Murphy, 'Climate Concern the Main Reason Voters Swung to Independents at Federal Election, Study Finds', *The Guardian*, November 2022.

195 Evans, *The Responsibility to Protect* (Brookings, 2008), p 101.

196 Ibid.

197 Ibid.

198 Some examples drawn from Japan's world-leading contributions to DRR. See for example, Goshi Tsukamoto, 'The Significance of International Disaster Reduction Cooperation in Japan's Diplomacy', Opinion Editorial, *Brookings*, 7 November 2012.

199 Ceridwen Dovey, '"It's a Worldwide Problem": How North American and Australian Firefighters Work Together', The New Yorker, 1 February 2020.

200 See for example, 'Early Recovery, Risk Reduction, and Resilience (ER4)', *USAid* (accessed 2022); 'Disaster Risk Reduction (DRR)', *Met Office* (accessed 2022).

201 Rajesh K. Mall, Ravindra K. Srivastava, Tirthankar Banerjee, Om Prakash Mishra, Diva Bhatt1, and Geetika Sonkar, 'Disaster Risk Reduction Including Climate Change Adaptation Over South Asia: Challenges and Ways Forward', *International Journal of Disaster Risk Science*, Vol 10, 2019, p 14–27.

202 'Members', *Coalition for Disaster Resilient Infrastructure* (accessed 2022).

203 'Our Mandate', *UN Office for Disaster Risk Reduction* (accessed 2022). See also, 'International Strategy for Disaster Reduction', ARES/56/195, General Assembly Resolution, 21 January 2002, p 2.

204 Evans, *The Responsibility to Protect* (Brookings, 2008), p 102.

205 See also, 'Arms Control Association: Treaties and Agreements' (accessed 2022).

206 '"This Could Be the COP Where We Lose 1.5C," Warns Alok Sharma – as It Happened', *The Guardian*, 15 November 2022. Mark Maslin, Priti Parikh, Richard Taylor, and Simon Chin-Yee, 'COP27 Will Be Remembered as a Failure – Here's What Went Wrong', *The Conversation*, 22 November 2022.

207 'ECOLEX: The Gateway to Environmental Law' (accessed 2022).

208 Erika Weinthal and Neda Zawahri, 'Don't Politicise Water: Why Water Cooperation Is Key for Israel, Jordan and the Palestinian Authority', *Foreign Policy Magazine*, 17 September 2020. See also, B. G. Verghese, 'Water Conflicts in South Asia', *Studies in Conflict and Terror*, Vol 20, 1997, 185–194.

209 Peter Newell and Andrew Simms, 'Towards a Fossil Fuel Non-proliferation Treaty', *Climate Policy*, Vol 20, No 8, 2020, p 1043–1054. See also, 'The Fossil Fuel Non-Proliferation Treaty' (accessed 2022).
210 Anthony Velazquez Abad and Paul E Dodds, 'Green Hydrogen Characterisation Initiatives: Definitions, Standards, Guarantees of Origin, and Challenges', *Energy Policy*, Vol 138, March 2020, p 1–13. See also, leading forest certification scheme the 'Forest Stewardship Council'.
211 'Arms Embargoes', Stockholm International Peace Research Institute (accessed 2022).
212 Evans, *The Responsibility to Protect* (Brookings, 2008), p 103 and 128.
213 Dual-use Export Controls', Stockholm International Peace Research Institute (accessed 2022). 'Dual Use Export Licenses', *US Department of Commerce* (accessed 2022).
214 'Export Controls: Dual-Use Items, Software and Technology, Goods for Torture and Radioactive Sources', *UK Government* (accessed 2022); 'Examples of [US] Dual Use Items', The University of Oklahoma (accessed 2022); 'Exporting Dual-Use Items', *European Commission* (accessed 2022). For a precise lists of dual use items, see: 'Commercial Control List', *US Department of Commerce*, 2022; 'Regulation (EU) 2021/821: Setting Up a Union Regime for the Control of Exports, Brokering, Technical Assistance, Transit And Transfer of Dual-Use Items', *European Union*, 2021; 'UK Strategic Export Control Lists: The Consolidated List of Strategic Military and Dual-Use Items that Require Export Authorisation from Great Britain and Northern Ireland', *UK Department of International Trade*, 2021.
215 'Select RAND Research on Security Cooperation, 2006–2019', *RAND*, 2021, p 1.
216 Ibid. See also 'Security Cooperation Overview', *US Defense Security Cooperation Agency* (accessed 2022).
217 Ibid.
218 Security Cooperation Overview', *US Defense Security Cooperation Agency* (accessed 2022). Export can also flow through the Direct Commercial Sale programme, Excess Defense Articles, leasing agreements, among other avenues.
219 See for example, *US Defense Security Cooperation Agency*, 2022: 'Access to Database', Excess Defense Articles; and 'C4 – Foreign Military Sales Program General Information: What May Be Purchased Using the FMS Program'. For country specific results search 'U.S. Security Cooperation' on *US State Department* website.
220 'U.S. Security Cooperation with Bangladesh', Fact Sheet, *US State Department*, 6 April 2022.
221 *The Responsibility to Protect*, ICISS, p 58. See also, Evans, *The Responsibility to Protect* (Brookings, 2008), p 103. See also, 'Gunboat Diplomacy', *National Museum of American Diplomacy* (accessed 2022).
222 '"I Want to Help My Own People": State Control and Civil Society in Burma after Cyclone Nargis', Report, *Human Rights Watch*, 2010, p 28.
223 Evans, *The Responsibility to Protect* (Brookings, 2008), p 102–103.
224 Ibid.
225 'United Nations Preventive Deployment Force (UNPREDEP)', Government of Canada (accessed 2022). See also, Thierry Tardy, 'United Nations Preventive Deployment Force (UNPREDEP—Macedonia)', in Joachim A. Koops, Thierry Tardy, Norrie MacQueen, Paul D. Williams (eds), *The Oxford Handbook of United Nations Peacekeeping Operations* (Oxford: Oxford University Press, 2014), p 500–510.

5 Climate Action and the Responsibility to React

> When prevention fails, conflict breaks out within a state, and mass atrocity crimes are occurring or imminent, it is not an option for the world to stand by and do nothing.
>
> (Gareth Evans, *The Responsibility to Protect*, 2008, p 105)

> We will do everything possible to mobilize the international community to support your country and to support all of you in this dramatic situation, in which beyond the numbers, I see families that lost their loved ones, their houses, their crops, their jobs, and are living in desperate conditions.[1]
>
> (UN Secretary-General, António Guterres, responding to widespread floods in Pakistan, September 2022)

This chapter parallels, and integrates where possible, climate action policies and ideas with the full range of strategies and policies under the Responsibility to Protect's (R2P) 'reaction' phase. The reaction phase, as with the prevention phase, comprises four broad sets of strategies (political, economic, legal, and security), with a fifth strategy, coercive military intervention, reserved for the most extreme cases. The strategies, and policy options therein, operate in a sequence moving from less to more intrusive and from less to more coercive.[2] The R2P agenda makes clear who, in practice, should do what, when, and how to halt or avert an atrocity situation that is occurring or imminent. The primary responsibility to react lies with the state itself where the crisis is erupting; but if the state is unable or unwilling to act, the responsibility to take appropriate action falls to the international community.[3]

For the climate action agenda, emergency responses are built in an ad hoc and uncoordinated way between individual states, UN agencies, regional organisations, and NGOs – often following a request by the affected country.[4] The human security agenda offers some level of conceptual coordination in terms of framing the relevant issues involved, but has little operational utility in answering the question as to who should do precisely what, and when, in response to a climate-related human protection catastrophe (see Chapter 3).[5] More promisingly, the EU, US, and UK have recently published crisis response options as a part of their push to integrate climate change into conflict

DOI: 10.4324/9780367201364-8

prevention and response policies. But this integrated agenda is still developing with only a few policy options proposed and little explanation offered on individual policies or how they might link together. The R2P reaction phase thus adds value to the climate action agenda in two significant ways: it indicates who should do what and when (the state itself, if not the international community, to stop human protection calamities); and how (by threatening, then implementing, if required, a combination of escalating conflict response policies).

Political and Diplomatic Strategies

Diplomatic Peacemaking

Diplomatic peacemaking, one of the less intrusive and coercive options along the Reaction spectrum, involves the early and effective mobilisation of political and diplomatic resources to bring back under control an explosive mass atrocity situation.[6] To help retrieve an out-of-control situation, governments can seek to mobilise one or a combination of four key resource domains: eminent individuals, the UN secretary-general and his 'good offices', regional organisations, and International NGOs (INGOs).[7] Their core mission, once deployed, is to try to persuade the key parties through negotiation, mediation, and facilitation that their interests and pride will not suffer from implementing a civilian protection agenda. To briefly address each resource domain in sequence.

Eminent individuals can play a range of interventionist roles from a very hands-on approach such as Richard Holbrook who 'cajoled and pummelled to a conclusion' the Dayton peace negations in the former Yugoslavia in 1995, through to a more hands-off approach such as the Kofi Annan-led AU-Mandated Panel of Eminent African Personalities that 'managed by their presence to calm the situation and eventually mediate a settlement' to the post-election violence in Kenya in 2008.[8] In the latter situation, Annan and his team built an agreement around redressing a number of historic and systemic grievances including establishing a plan for ongoing negotiations on the underlying root causes of the violence, especially those related to the constitution and land allocation and use. The Constitution of Kenya, the key outcome of the negotiations, was the genesis of a process that saw Kenya elevate climate action to a core government priority and subsequently become a continental leader on climate change.

The UN secretary-general, in the exercise of his 'good offices' role, has dozens of special representatives, personal representatives, special envoys, and special advisers working on peace and security issues, heading field-based missions, at headquarters, or acting as roving envoys.[9] The Secretary-General António Guterres has delivered the strongest messages of any global leader on the importance of climate action, and the Department of Political and Peacebuilding Affairs (DPPA), which houses the 'good offices' functions, has begun to integrate climate-informed personnel into their peace engagement activities worldwide. For example, in 2020 a Climate Security Advisor was appointed to the UN Assistance Mission in Somalia, and a Climate, Peace, and Security Advisor will soon be appointed to the UN Assistance Mission for Iraq.[10]

Regional organisations across the world, not only those representing countries in climate hotspots such as the African Union (AU) or Association of Southeast Asian Nations (ASEAN), but NATO, the EU and elsewhere, recognise the serious threat that climate change poses to peace and security, and have begun to develop strong climate-informed peace engagement activities.[11] In addition, INGOs involved in peace diplomacy such as the Crisis Management Initiative (CMI),[12] The Carter Centre,[13] and The Centre for Humanitarian Dialogue (HD)[14] are increasingly building climate action issues into their operations and advocacy. The Elders, a group of 15 eminent global leaders, many of whom have been involved in peacemaking initiatives, have almost all spoken out about the threat posed by climate change and the need for sustained climate action to protect vulnerable populations.[15] For example, Former President of Finland Martti Ahtisaari – who has played a highly important role in mediating peace agreements over decades in his capacity as UN representative, EU representative, and head of his own NGO the CMI[16] – has explained that 'if emissions, temperatures and sea levels continue to rise, large parts of Earth could become literally uninhabitable. I believe this would be an unconscionable betrayal of future generations'.[17]

Political Incentives

Peace negotiators and those trying to encourage an improvement in a country's behaviour towards their civilian population can sometimes offer political incentives including diplomatic recognition, membership to an international organisation, and military assistance; as well as those orientated towards economic options such as development aid.[18] Rendering this option climate-relevant could in a general sense involve peace negotiators and practitioners expanding their mission scope beyond traditional human rights concerns to include climate change. For example, by deploying political incentives to persuade a government to provide enhanced protection services to its population from an occurring or imminent climate-related harm. To achieve this, the political incentives identified above could be adjusted to reflect a climate change character in the following way.

First, formal diplomatic recognition of a government, or the lesser option of re-establishing diplomatic relations with a country (the difference is, for example, the US recognises the government of Iran, but has no formal diplomatic relations with it),[19] could be in part premised on the target government enacting domestic climate action reforms. For example, in November 2021, as a core part of its diplomatic and economic reengagement with Syria, the

United Arab Emirates signed a deal to construct a large solar power plant near Damascus.[20] Such a deal for Western countries would rightly need to involve the much larger task of establishing a process for a political transition to occur in Syria.[21] Second, in many instances, membership to an international organisation does not necessarily mean that the country in question can fully participate in the policymaking process occurring within the organisation. For example, the extremely complex interrelated issues negotiated at UNFCCC gatherings can be successfully navigated by the teams of experts from developed countries, but have proven more challenging to grasp for some resource-poor developing countries – which in general have the most at stake from the outcome.[22] Western governments have an obligation to help empower developing countries in the international climate negotiations under Article 6 of the 1992 Framework Convention on Climate Change and under Article 12 of the 2015 Paris Agreement.[23] Support can include assistance with climate impact assessments, greenhouse gas inventories, and UNFCCC and climate finance processes. Going beyond what is required could be premised on the realisation of human protection goals in the recipient country – in this way, acting like a form of aid conditionality. Third, the opportunity to acquire climate-related 'dual-use goods' such as radar and weather monitoring systems, heavy machinery and equipment for natural hazard reduction and disaster recovery such as helicopters and firefighting vehicles (previously discussed), could help incentivise a rapid behavioural change in the target country.

Political Sanctions

Political sanctions can include the suspension or expulsion of a country from a regional or international organisation; exclusion from a summit or meeting; the withdrawal of diplomatic recognition; and/or the application of bilateral diplomatic restrictions and expulsions. They can also encompass targeted travel and visa bans against specific political leaders, their business associates, and family members. Cultural isolation such as sports, arts, and academic boycotts, and 'naming and shaming' through condemnation in international forums also fall within the remit of political sanctions.[24]

Western governments could explore the following four options to encourage governments to pursue policies more conducive to the protection of civilians. First, work to suspend or expel egregious human rights abusers from international or regional bodies, which may not only entail a loss of national prestige, but also a loss of financial assistance and technical cooperation on climate change. Second, government decision-makers responsible for climate change and/or energy and disaster management, and their families, could be subjected to targeted travel bans and visa bans; while those family members attending Western schools and universities could have their certifications and degrees suspended. Third, cultural sanctions, which can be effective in raising awareness internationally and in the target country that a particular behaviour is unacceptable, can take three distinct forms: Western academic institutions could enact boycotts involving the restriction of research funding or publishing opportunities, or cancelling academic and student exchanges; Western artists and performers could be encouraged to reconsider tours to the target country or its region, and target country artists and performers could be banned from tours in Western countries; sporting boycotts could also be enacted, which have over the years involved expulsions from the Olympics and international soccer tournaments, and the cancellation of cricket and rugby tours.[25] Fourth, 'naming and shaming' has been a staple option pursued by NGOs and countries at international climate events for the past 30 years. This option tends to be largely ineffective when pursued by NGOs, evidenced by the Climate Action Network's repeated high-profile 'fossil of day' awards delivered to Australia, the US and Canada during UNFCCC negotiations; but when delivered by political leaders, naming and shaming can make a difference.[26] For example, Australia increased its national emissions reduction target from 5 to 26–28 per cent (in line with the US target) after US President Barrack Obama delivered a rousing speech at the University of Queensland in 2014 calling on countries to 'step-up' on climate action – clearly directed towards Australia itself.[27]

Economic Strategies

Economic Incentives

Promising (or delivering) economic incentives can contribute to resolving a conflict situation, for instance, by persuade parties to negotiate a peace deal, or to achieve a longer-term strategic goal such as relinquishing nuclear weapons.[28] Measures can include concessions on trade access, development assistance, beneficial trade arrangements, investment offers, or membership in a regional organisation.[29] Positive economic incentives could also help sway a target government towards adopting climate action strategies for human protection purposes. For instance, economic incentives could be used in the immediate aftermath of a major climate-related human protection situation to persuade a government that was deliberately neglecting its own people to allow for the effective delivery of foreign aid, or at least to get them to the negotiating table – as was the situation in Myanmar in the days and weeks after Cyclone Nargis in 2008.[30] Western largesse, which could take the form of greater access to low-emissions technology or financial assistance, offered to induce behavioural change in emergency situations could also be deployed with longer-term climate goals in mind, for instance, to encourage the target state to shift to low-emission energy sources.[31]

Economic Sanctions

Economic sanctions can include trade restrictions (the refusal to export – otherwise known as 'embargoes'; and/or the refusal to import – 'boycotts');

covert refusals to trade ('blacklisting'); deprivation of ownership ('expropriation'), punitive taxation, aid suspensions, and asset freezes.[32] They can be comprehensive such as an embargo applied to an entire country, or they can be targeted such as blocking financial transactions with a particular business, group or individual.[33] The aim, no matter whether comprehensive or targeted, is to alter the strategies and policies of the target country, and are used by governments worldwide to elicit change on a range of topics of international concern including terrorism, narcotics, weapons proliferation, human rights, conflict resolution, and cybersecurity.[34] Criticisms include their dubious record of effectively altering target government behaviour, and their potential to deny essential goods and services to already vulnerable communities.[35]

Different forms of economic sanctions have been applied to climate-related activities such as on the energy sector including oil, gas and coal and related equipment and technology; the production and import of iron, steel, and other metals; and other critical infrastructure and industrial machinery.[36] Western powers have a long history of applying economic sanctions to fossil fuel products. Since the 1980s, the US has applied comprehensive economic sanctions on the National Iranian Oil Company and associated oil-sector activities (e.g. investment, shipping, insurance, and exports).[37] US oil-sector sanctions have also been applied to Libya's national oil company, which has had a considerable impact on the Libyan economy.[38] In addition, since 2014, targeted economic sanctions have been applied to Russian oil companies and activities (e.g. accessing debt finance, and oil exploration technology).[39] And in 2019, the US 'backlisted' Venezuela's national oil company, Petróleos de Venezuela, serving to eliminate the oil trade between the US and Venezuela.[40] US economic and financial sanctions originate from the executive or legislative branch – typically the President will issue an Executive Order to start the process.[41] Similarly, the EU has applied trade measures (embargoes) and financial measures (bans on bank loans) to limit export and investment in North Korean and Russian oil.[42] And has banned the importation of Iranian and Syrian oil into the EU zone.[43] EU economic sanctions form a part of its Common Foreign and Security Policy, and must be agreed to unanimously by its member states before being applied.[44] Individual EU countries can also impose sanctions independently. For example, in 2020, appalled by forest loss and wildfires in the Amazon, six European countries (Germany, France, the Netherlands, Norway, Denmark, and Italy) and the UK threatened Brazil with a trade boycott.[45]

Western governments could consider developing a 'climate-informed sanctions regime' that targeted extreme fossil fuel use and high emitting individual activities, with the aim to persuade countries to adopt low-pollution and climate-resilient development pathways. Such a regime may prove effective in immediate human protection situations because in many conflict-affected and fragile states such as Syria and Myanmar, major industry is generally owned and operated by senior government and military officials.

Legal Strategies

Criminal Prosecution

Actively seeking to arrest, try, convict, and punish an individual threatening or perpetrating atrocity crimes on a population is best pursued, with or without international assistance, through the target state's own criminal justice system.[46] But if this is unachievable, which could be the case for several reasons including the unwillingness of the target state to cooperate, then the three main international legal options to consider are special tribunals, the International Criminal Court, and national courts able and willing to exercise 'universal jurisdiction' (see Chapter 4).[47]

Climate change litigation is rapidly increasing around the world. According to the UN Environment Programme (UNEP), climate litigation cases nearly doubled between 2017 and 2020 – from 884 cases brought in 24 countries to 1550 cases filed in 38 countries.[48] One of the recent key trends in successful climate litigation has been termed 'climate rights', which asserts that the defendant (usually a government or corporation) has pursued insufficient actions to mitigate climate change and therefore violated the plaintiffs' (usually an individual or community – represented by a civil society organisation) constitutional and international rights to life, health, food, water, liberty, family life, and more.[49]

Applying this logic, domestic courts have ruled in several high-profile cases that governments do have a duty of care to protect their citizens from climate impacts by pursuing, for instance, more aggressive emissions reduction targets in line with the Paris Agreement's temperature goals. The most prominent example occurred in 2015 when the Dutch Supreme Court ruled that the Dutch government's emissions reduction target of 17 per cent below 1990 levels by 2020 was an insufficient contribution to achieve a safe climate for Dutch citizens and ordered an increase to 25 per cent.[50] This ruling was upheld in 2019, confirming that the government had a duty of care to its citizens in accordance with its obligations under the European Convention for the Protection of Human Rights and Fundamental Freedoms (which had been integrated into Dutch law).[51]

Similar 'climate rights' cases have been brought against central governments (in Canada 2018, Peru 2019, United States 2020, Canada 2020, Brazil 2020, and South Korea 2020), environmental ministries (in Colombia 2018), ministers (in Australia 2021), specific government policies (in Philippines 2017, Pakistan 2019, and Austria 2020), and specific government projects (in Norway 2020).[52] A clear limitation is implementation. For example, in the case of *Future Generations* vs *Ministry of the Environment* ('Colombia 2018' above), the Supreme Court of Colombia found that deforestation in the Amazon threatened the plaintiffs' 'climate rights' enshrined in the national constitution and ordered the government to develop and implement a plan to halt deforestation.[53] Analysis a year later found that none of the specific actions that the court ordered had been fully undertaken.[54]

Tribunals have also seen climate rights cases. A key case occurred in 2014 where India's National Green Tribunal found that vehicle emissions contributed significantly to glacial retreat in the Himalayan region which violated Article 48A of India's Constitution (which requires the state to protect and improve the environment), Article 51A (requiring a citizen to protect and improve the natural environment) and Article 21 (protecting the right to life as a fundamental right).[55] The Tribunal ordered the provincial government of Himachal Pradesh (situated in the Western Himalayas) to undertake sweeping measures to reduce pollution such as random pollution checks, restricting transport in certain areas, and implementing a reforestation programme.[56] The UNEP's first report on Environmental Courts and Tribunals (ECTs) worldwide, published in 2016, observed an 'explosion' in the number of ECTs from only a handful in 1970 – mostly in Europe – to 350 in a variety of countries in 2009, growing to 1,200 operating in at least 44 countries by 2016.[57] The updated UNEP report on ECTs published in 2022 found 'steady growth' since 2016, with the number of operational ECTs now standing at 2,116 in 67 countries.[58] The UNEP has identified the 'recognition of the linkage between human rights and environmental protection' as one of the key drivers behind the growth of specialised ECTs.[59]

International legal fora have also attracted an increase in cases that seek to punish governments for failing to develop and implement adaptation and mitigation strategies consistent with the protection of international rights. For example, in May 2019 a group of eight Torres Strait Islander Peoples, Australian nationals, and six of their children submitted a petition against the Australian government to the United Nations Human Rights Committee, alleging that Australia had violated their fundamental human rights under the UN International Covenant on Civil and Political Rights (ICCPR) – Article 27 (the right to culture), Article 17 (the right to be free from arbitrary interference with privacy, family, and home), and Article 6 (the right to life) – by failing to, inter alia, upgrade seawalls on the islands and reduce national greenhouse gas emissions. In September 2022, the Committee found that Australia's failure to take timely and adequate measures to safeguard Torres Strait Islander Peoples from adverse climate change impacts led to the violation of their right to culture (ICCPR Article 27) and the right to be free from arbitrary interference with privacy, family, and home (Article 17), and asked Australia consider compensation, deep consultation, and other measures to secure the communities' safe existence on their respective islands.[60] Other 'climate rights' petitions and complaints have been filed under the United Nations Convention on the Rights of the Child and to UN Special Rapporteurs respectively.[61] The UNEP expects more climate rights cases to be brought before international adjudication bodies because, among other reasons, international fora offers strategic opportunities that may be unavailable for plaintiffs in national and subnational courts – for instance, domestic courts may be unwilling or ineffectual when seeking to hold a national government accountable.[62]

Western governments could consider providing legal support to individuals and communities located in fragile states (as well as those displaced across borders) whose 'climate rights' are being, or have been, negatively impacted by climate change and environmental damage. Other options include providing financial and technical assistance to develop legal forums that hear climate rights cases, and/or assist local actors implement court rulings.

Security Strategies

Military Strategies Short of Applying Coercive Force

Deploy Peacekeepers

Peacekeeping, in its traditional form, involves 'blue-helmeted' forces being engaged in the monitoring, supervision, and verification of ceasefires and early-stage peace agreements.[63] Peacekeeping missions are multilateral in character, generally operate under UN command (although regional organisations such as NATO or the AU have sometimes lead individually or jointly with the UN), and should be authorised by the UN Security Council under Chapter VI of the UN Charter (authorisation is implicit because the Charter does not explicitly mention peacekeeping).[64] All parties involved in the conflict are required to give their consent to the mission. Mission actors are expected to remain impartial and are not mandated to use lethal force unless in self-defence.[65] A more modern take on UN peacekeeping, termed 'coercive protection' or 'peacekeeping plus', involves UN military units (troops or police) being permitted through an expanded mandate to use lethal force not only to defend themselves but also to respond to peace 'spoilers' and to stop violence against civilians wherever they witness it (compared to simply watching and reporting).[66] Such expanded mandates require Chapter VII authorisation. Both traditional and modern operations, as suggested, are geared towards supporting established peace agreements or those under negotiation. Peacekeeping as a policy option becomes less relevant in situations where there is no peace agreement in sight – in other words, where the priority need is to create peace.[67]

About 80 per cent of UN-led peace operations (encompassing peacekeeping and political and peacebuilding missions) are deployed in countries highly exposed to climate impacts (see also Chapter 3).[68] All the largest African missions are in climate hotspots, including South Sudan, Mali, the Democratic Republic of the Congo, Central African Republic (CAR), Sudan, and Somalia.[69] The impacts of climate change have several implications for peacekeeping missions.[70] For example, slow onset events such as drought and rapid onset events such as floods, can cause food and water shortages, reduce economic opportunities, precipitate displacement, and stretch local and national government services with the potential to undermine their legitimacy.[71] Local armed groups and peace 'spoilers' can exploit these situations, individually or in combination, to further their own interests. In addition, climate impacts can reduce the mobility and performance of peacekeepers.

Much emphasis has been placed on the need for peacekeeping missions to reduce their environmental footprint.[72] However, UN peacekeepers have been involved in a variety of climate-related human protection activities beyond this. For example, UN missions in CAR (MINUSCA) and South Sudan (UNMISS), which are mandated to consider climate impacts in assessment and planning, have identified potential hotspots for resource-related violence on transhumance migration routes, and pre-emptively deployed military patrols to defuse tensions.[73] However, such initiatives remain ad hoc – often dependent on the priorities of the mission leader and the availability of dedicated climate specialists.[74] In another example, peacekeeping operations in Mali (MINUSMA) have supported climate-informed dialogue initiatives between farmers and herders to prevent violence on transhumance migration routes.[75] Other peacekeeping operations have engaged in climate-informed Security Sector Reform, Disarmament, Demobilisation and Reintegration, as well as securing critical environmental infrastructure such as water wells (UNAMID – Darfur).[76]

To address some of these challenges and to build associated mission capacity, in December 2021, the UN Department of Peace Operations (DPO) joined the Climate Security Mechanism (CSM; see Chapter 3).[77] Drawing on the UN-wide expertise on offer under the CSM, the DPO plans to further integrate climate-related concerns into its mandated mission work, tools, and staff training, including by deploying climate security advisors to missions in climate hotspots (initially UNSOM – Somalia and UNAMI – Iraq).[78] Western governments could continue to advocate for UN peacekeeping missions to systematically integrate climate-related concerns and issues into their mission mandates and activities related to human protection (e.g. risk assessments and environmental peacebuilding).

Establish Safe Havens

A safe haven (a term that can be used interchangeably with 'safe area' and 'safe zone') is a low-intensity military operation undertaken by international actors within a target state's borders to shelter and protect, within a temporary and designated geographical area, civilian populations and internally displaced persons (IDPs) facing an imminent crisis or threat.[79] The advantage of establishing a safe haven is that it allows for a third party to intervene in a conflict where a full peacekeeping mission is politically impossible and nevertheless achieve civilian protection.[80] Ideally however, safe havens should be agreed upon by all parties to a conflict and enforced with multilateral authorisation.[81] It cannot be assumed that a safe haven will be automatically respected by hostile forces hell-bent on doing harm to a population. Therefore, a safe haven is only as good as the mandate and capability of the intervening force to defend it, with deadly force if necessary, for the duration of the safe haven's existence.[82] The presence of a safe corridor to the safe haven that guarantees humanitarian access increases the chances of successfully protecting civilians.[83] Done right, safe havens can save thousands of people that might have otherwise been killed.

Safe havens of sorts, which have occasionally involved a military presence, have long been used to protect environmental assets and wildlife from human activities such as poaching, logging, and the spread of invasive species.[84] In the climate change discourse, the notion of 'safe havens' have been used to describe parcels of forested land in developing countries to be protected in order to offset emissions from industrial processes in developed countries;[85] to prescribe the need to protect and regenerate ecologically sensitive areas such as coasts and hill-slops for Disaster Risk Reduction purposes;[86] and in popular debate, to identify the safest location globally to avoid the worst impacts of climate change.[87]

An integrated policy response could involve Western governments establishing a traditional military safe haven in a foreign country to protect populations who have been displaced by a climate-related hazard. Groups that have been displaced due to climate factors may be at enhanced risk of direct attack from hostile parties in the country in question, or at heightened risk of severe malnutrition and starvation as a result of acute food and water shortages in climate hotspots. Establishing a safe haven to protect refugees and IDPs has strong precedent. For example, Operation Provide Comfort in 1991 involved the UK and US, with Türkiye, deploying 17,000 troops over an area of 21,000 km^2 in Northern Iraq in order to safeguard an estimated 450,000 Kurdish refugees.[88] According to the UN High Commissioner for Refugees, climate-related hazards are already causing more than 20 million people to leave their homes and move to other areas in their countries each year, with vulnerable people living in some of the most fragile and conflict-affected countries disproportionately affected.[89] Forecasts of the number of people facing displacement and migration because of combined rapid and slow-onset climate hazards vary from about 200 million people by 2050 to more than 1 billion.[90]

Establish No-Fly Zones

No-Fly Zones (NFZ) are the aerial equivalent of safe havens and involve an outside actor or actors prohibiting some or all aircraft flight over a specified territory and undertakes measures to intercept aircraft violating the prohibition or otherwise punish those responsible for violations.[91] Key policy decisions include the type of aircraft to be restricted and where, and what enforcement mechanisms and broader penalties will apply for NFZ violation.[92] Above all however, the effectiveness of NFZs as a credible civilian protection option (as with safe zones) requires an operational capability and willingness to apply full-scale military force in their defence, if that proves necessary.[93]

Aerial assets have long been deployed to deal with climate-related issues. Helicopters are used in search and rescue operations in the aftermath of a rapid-onset event such as a cyclone, and also contribute significantly to the movement of essential emergency medical, food, and water supplies around a country. For example, in response to Typhoon Haiyan that struck the Philippines in 2013 – one of the most powerful and deadly tropical cyclones ever recorded with winds above 315 km/h and killing more than 6,500 people – the US military moored USS George Washington off the Philippine coast to provide a platform for helicopters to move emergency medical, food, and water supplies and perform a search and rescue role – the US's presence tripled the number of helicopters available.[94] In addition, fixed-wing aircraft are routinely used to drop food and medical supplies in regions suffering from slow-onset events such as protracted drought. For example, the UN World Food Programme working with INGOs Mercy Corps and the International Red Cross regularly fly large fixed-wing aeroplanes over a number of drought-impacted regions in North Africa, dropping dozens of individual pallets stacked with emergency food supplies to drought-affected areas.[95] While airdrops can be expensive, and are most effective when coordinated with ground forces, they can be nonetheless essential to the immediate survival prospects of populations located in inaccessible (perhaps due to terrain and/or conflict) climate-ravaged regions.[96] Other relevant uses of aerial assets in climate adaptation and mitigation strategies include aerial firefighting and surveillance of environmental assets (i.e. illegal logging of carbon stocks). However, it is quite a significant step to argue that a coalition of foreign states should use military assets to establish and enforce a NFZ in a sovereign country to protect civilians from climate-related harm.

An instance that may attract a conventionally motivated NFZ to be established could involve situations where a hostile government is using air power to destroy 'environmental infrastructure' (e.g. water resources, sanitation and waste facilities, and energy infrastructure) critical to sustain human life and livelihoods. Avoiding energy infrastructure is especially important because the production and distribution of water at scale in most countries relies on a functioning electricity network. Sabotaging powerplants and transmission lines therefore effectively shuts down water pumping stations, water treatment plants, and sewage treatment plants. Aerial bombardment of 'environmental infrastructure' has become a normalised tactic in recent conflicts in the Middle East and North Africa including in Sudan, Libya, Syria, and Yemen.[97] These regions and countries are also considered climate 'hotspots', all suffering from severe drought. For example, as a part of its campaign of mass murder in Darfur between 2003 and 2005, the Government of Sudan used air power (Antonov airplanes and helicopter gunships) to not only directly bomb civilians but also blow-up water wells, effectively making the area uninhabitable.[98] Western countries have also employed this tactic. For example, in Libya in 2011, while armed groups destroyed water wells from the ground, resulting in reduced water supply to civilians, NATO bombed Gaddafi's 'Great Man-Made River' from the air causing major disruptions to clean and safe drinking water, sanitation and hygiene throughout Libya.[99]

Establishing a 'climate-informed no-fly zone' to stop the deliberate targeting of environmental infrastructure critical to sustain life in a climate hotspot could be pursued in the following way. The legal instruments include Protocols I and II to the 1976 Convention on the Prohibition of Military or Any Other Hostile Use of Environmental Modification Techniques.[100] As part of the basic rules pertaining to warfare, Article 35 prohibits 'methods or means of warfare which are intended, or may be expected, to cause widespread, long-term, and severe damage to the natural environment'; while Article 54 states that 'it is prohibited to attack, destroy, remove or render useless objects indispensable to the survival of the civilian population', which includes civilian infrastructure such as 'drinking water installations and supplies and irrigation networks'. Militaries are to avoid attacking such installations so as not 'to leave the civilian population with such inadequate food or water as to cause starvation or force its movement'.[101] It should also be noted that since the 1990s, the UK and the US have established NFZs in several conflict zones with and without legal cover from the Security Council. For example, a NFZ was established in Bosnia in 1993 with Security Council authorisation, while in Iraq in 1991 and 2003 without authorisation.

Implement Arms Embargoes

An elevated option from threatening an arms embargo as a preventive measure (see Chapter 4) is to implement one should a conflict break out and atrocities start to occur. Arms embargoes can be extended to encompass 'dual-use goods' used for climate adaptation and mitigation activities ranging from the small, such as weather monitoring computer equipment, to the large such as search and rescue helicopters and related equipment. The UK, US, and EU can apply sanctions on dual-use goods individually and collectively through various mechanisms – and/or under Chapter VII authority of the UN Security Council – and have done so on several countries in reaction to human protection violations.

Restrict Communications

Attempting to shape (restrict or otherwise) social media content (and other communications such as radio and mobile phone frequencies) can play an important role in protecting civilians during a crisis.[102] The most notorious example of broadcast media being used to enable mass atrocities occurred in Rwanda in 1994 with the Hutu extremist station, *Radio Mille Collines*, priming then helping direct the genocide. Nowadays, social media platforms especially – like newspaper, radio, and television broadcasts before it – can undermine efforts to prevent, manage, and resolve a crisis. In 2021, there were approximately 4.2 billion active social media users – 53 per cent of the global population.[103] On average, social media users have accounts on eight different platforms and spend two-and-a-half hours per day on social media. Most users are between 17 and 24 years old.[104] Consequently, according to the DPPA, 'social media increasingly shapes political and social interactions and, indeed, people's perceptions of reality'.[105] Threats, hate speech and the intentional creation and circulation of mis-and-dis information about the situation and key stakeholders are becoming commonplace in crisis situations as was seen prior to, during and after, both the so-called 'clearance operations' in Myanmar in 2017[106] and the Libyan peace process in 2020.[107]

Western countries could consider working with social media companies to establish processes and programmes that enable the accurate conveyance of information through social media (and, if possible, traditional broadcast and print media) in the aftermath of a climate-related human protection catastrophe. Options could include, helping to establish in-country social media monitoring services with staff knowledgeable about climate impacts and opportunities; combating misinformation and disinformation on social media platforms about climate impacts and responses, such as the intention of aid interventions, by identifying such information and providing links to genuine information such as official UN statements; and by establishing processes that would allow for a 'digital ceasefire' between key stakeholders (and their constituencies) to be implemented should an environmental emergency occur.[108]

Establishing these processes and programmes could possibly help avoid repeating the problems faced in the aftermath of Cyclone Nargis. In this situation, extreme winds and storm surge cut electricity lines and destroyed telecommunication networks across the Irrawaddy Delta (Southwest Myanmar).[109] State-controlled media (falsely) portrayed the government as fully in control of the situation with extensive coverage of the military clearing away debris and distributing aid.[110] Very little coverage captured the full extent of the devastation across the Delta or the immense loss of life.[111] Local independent journalists seeking to overtly cover the disaster were arrested.[112] Foreign emergency relief and assistance were framed as threatening. Foreign journalists were banned from entering the country, and those already in the country, and caught in the Delta, were deported.[113] It was not until two weeks after the cyclone had struck – with the death toll climbing towards the final figure of 130,000 with 2.4 million displaced – that the UN agency for information and communication technologies (the International Telecommunication Union) was able to deploy satellites to the Delta to help restore communication links for government officials and the few humanitarian agencies on the ground.[114]

Military Intervention

The use of coercive military force has dominated academic and popular discussions about R2P (much of it is highly critical). Gareth Evans calls this focus 'unfortunate' because it has led many to 'misunderstand R2P as being *only* about the use of military force', when in fact 'it is about much more than that – about prevention at least as much as, if not more than, reaction, and about many much less extreme kinds of reaction' (as Part II of this book also demonstrates).[115] But, he continues – pointing to Rwanda in 1994 and Srebrenica in 1995 as examples where military intervention would have saved many thousands of innocent lives – 'if there is one thing as bad as using military force when we should not, it is *not* using military force when we *should*'.[116]

The ICISS made clear that military force should be used only in a very limited range of cases, not only where it was legal under international law, but where it satisfied a number of additional prudential criteria (see Box 1). Although such criteria were never formally adopted, either in the original World Summit resolution or by subsequent Security Council decision, they remain an important part of any contemporary discussion of the possible use of force in response to mass atrocity crimes.[117]

Box 1. The Use of Force: Criteria of Legitimacy[118]

One: Seriousness of Harm

Is the threatened harm of a kind, and sufficiently clear and serious, to justify *prima facie* the use of military force? In the case of internal threats, does it involve genocide, war crimes, ethnic cleansing or crimes against humanity, actual or imminently apprehended?

Two: Proper Purpose

Is it clear that the primary purpose of the proposed military action is to halt or avert the threat in question, whatever other purposes or motives may be involved?

Three: Last Resort

Has every nonmilitary option for meeting the threat in question been explored, with reasonable grounds for believing that other measures will not succeed?

Four: Proportional Means

Are the scale, duration and intensity of the proposed military action the minimum necessary to halt or avert the threat in question?

Five: Balance of Consequences

Is there a reasonable chance of the military action being successful in meeting the threat in question, with the consequences of action not likely to be worse than the consequences of inaction?

The notion that coercive military force could be used to achieve climate-related human protection goals – for example, to protect human populations in the aftermath of a sudden onset event (e.g. cyclone, flood, or wildfire), or a population in grave danger from a slow onset event (e.g. extreme drought) – have conceptual anchors in 'ecological intervention' and 'eco-humanitarian intervention'.[119] The former involves responding militarily to major environmental emergencies that could have grave transboundary consequences for human populations (e.g. a Chernobyl-style nuclear explosion). The latter in contrast, involves military intervention to prevent or halt deliberate and extreme damage to environmental assets, otherwise known as 'ecocide', critical to sustain human life in that country (e.g. water resources), irrespective of the spill-over effect.[120] A third category, and the least likely to garner international support, involves military intervention to prevent or halt the large-scale destruction of high-value environmental assets (e.g. species close to extinction or rainforests) – again, an 'ecocide' situation – for their own sake, that is, with no serious human rights violations occurring.[121]

These ideas were tested in Brazil in 2019. In this scenario, the government of Brazil, led by President Jair Bolsonaro, embarked on a development agenda in the Amazon rainforest that involved burning and clearing large swathes of land for agriculture, mining, and other industrial enterprises. At one point in 2019, the world's largest rainforest, 60 per cent of which is located in Brazil, was engulfed in nearly 72,000 fire outbreaks.[122] The fires produced significant qualities of carbon dioxide, destroyed biodiversity, and involved small-scale human rights abuses against indigenous populations opposed to the policies.[123] As the fires burnt out of control, openly encouraged by Bolsonaro, French President Emanual Macron made a series of remarks in 2019 that indicated that international intervention was a possibility should Brazil fail in its responsibility to protect the Amazon. For example, and touching on elements of the above typology, Macron described the fires as an 'ecocide' situation, and one that involves France because 'France's biggest exterior border is between Guiana and Brazil', and also 'an international crisis' because the Amazon 'produces 20 percent of our planet's oxygen'.[124] Macron subsequently persuaded the G7 members to commit US$20 million to Brazil to finance firefighting aircraft.[125] President Bolsonaro's response was to claim that Macron and other G7 members had a 'colonialist mindset' and rejected the funding offer.[126] This high-level interaction was enough to trigger a discussion about the role that R2P's reaction phase could play in the immediate protection of the Amazon. For example, eminent International Relations Professor, Stephen Walt, explicitly brought the R2P into the discussion, and wrote that

> It's only a matter of time until major powers try to stop climate change by any means necessary...Brazil isn't a true great power and threatening it with either economic sanctions or even the use of force if it refused to protect the rainforest might be feasible.[127]

Similarly, Professor of Law, Lawrence Douglas, wrote that 'Massive environmental crimes, such as those presently unfolding in the Amazon, necessarily have a spill-over effect, as the degradation of the rainforest will do grave, and arguably irreversible, damage to our planet's climate…[and the R2P] offers a sound template for a workable practice of environmental intervention' both in terms of its principles and graduating policy framework from trade sanctions through to 'deploying military force'.[128]

Owing to the immense political, intellectual, and substantive complexities involved in the debate about external military intervention for humanitarian and/or environmental reasons (e.g. competing notions of 'legitimacy' and what constitutes a 'just war'), this final section of Chapter 5 does not seek to detail the precise circumstances in which the use of force becomes a live option for decision-makers, but rather simply identifies the practical international legal avenues available to decision-makers when seeking a coercive military solution in response to an imminent or occurring human protection catastrophe.

Security Council Authorisation under Chapter VII

Chapter VII of the UN Charter is the international locus for the authorisation of coercive military force. The important provisions are Article 39, which says that 'The Security Council shall determine the existence of any threat to the peace, breach of the peace, or act of aggression and shall...decide what measures shall be taken...to maintain or restore international peace and security'; and Article 42, which says that

> Should the Security Council consider that measures provided for in Article 41 [less coercive measures] would be inadequate or have proved to be inadequate, it may take such action by air, sea, or land forces as may be necessary to maintain or restore international peace and security.[129]

These provisions provide clear guidance on the responses available for traditional cross-border threats to peace and security. However, for internal human protection catastrophes – as the R2P framework is primarily geared towards addressing – their application becomes less obvious.[130] In such situations, what matters most is the Security Council's interpretation of what constitutes an 'international threat to peace and security'. As Evans explains: 'with no higher authority to gainsay it, threats to international peace and security are what the Security Council says they are' and since the end of the Cold War, it has defined them quite broadly as it did in Bosnia and Somalia.[131] However, this latitude in defining threats to peace and security has not been extended to the climate action agenda.

As Chapters 2 and 3 showed, since 2007 Western governments and successive UN Secretary-Generals' have formally advocated for the Security Council to recognise climate change as a threat to peace and security,[132] with increased pressure applied to the Council since the Paris Agreement was reached in 2015.

The most recent attempts occurred in September and December 2021.[133] But as with all previous efforts, China and Russia opposed (abstaining and vetoing respectively in this instance) the Council even adding climate change to its formal agenda – let alone defining climate change as a threat to international peace and security. They did this based on, first, an unconfirmed relationship between climate change and security, and second, the UNFCCC being the most appropriate site to address climate issues.[134] In response, the US Ambassador to the UN, Linda Thomas-Greenfield, asserted that by using its veto Russia had 'stopped the world's most important body for maintaining international peace and security from taking a small, practical and necessary step to combat the impacts of climate change'.[135] Whether it is a small step or not, there is a growing consensus among UN member states,[136] and the INGO community that specialises in conflict prevention and response,[137] that the Council needs to begin to formally engage with climate security issues as a matter of course.

Self-Defence under Article 51 (Chapter VII)

The use of military force under Chapter VII does not always require Security Council authorisation. Article 51 makes clear that 'Nothing in the present Charter shall impair the inherent right of individual or collective self-defence if an armed attack occurs against a Member of the United Nations'.[138] In other words, Article 51 shifts the decision to use force from the collective (the Security Council) to the individual state.[139] This shift has been the catalyst for states to repeatedly seek to expand the scope of what constitutes legal coercive action under Article 51. The sequential US-led wars in Afghanistan and Iraq are the most prominent examples of this expansionist agenda.[140] To explain, in the immediate aftermath of the attacks on the Twin Towers in September 2001, the US argued that the scope of Article 51 should be extended to encompass the use of force against non-state actors such as terrorist organisations as well as 'those responsible for aiding, supporting or harboring' terrorists.[141] On this the Security Council agreed, which paved the way for the invasion of Afghanistan. In 2002 and 2003, the US again appealed to Article 51 to justify a military invasion of Iraq. The orthodox idea governing the right to self-defence – and therefore the use of defensive military force – as set out in the Charter, requires the occurrence of an 'armed attack'. Here however, the case was made that Iraq's Weapons of Mass Destruction stockpiles (proven to be false) constituted an imminent threat to the US which required an anticipatory military response. Evans explains that the claim to anticipatory defence, which was far more reaching than any previous claims, stretched Article 51 beyond endurance.[142] What the US's actions show is that the scope of Article 51 – in terms of what constitutes a threat (non-state actors and those protecting them in the first instance) and what actions states can take to neutralise a threat (pre-emptive action in the second instance) – is fluid and open to redefinition.

Western governments and their security agencies overwhelmingly agree that threats of major concern no longer simply involve 'armed attack' undertaken by states, but rather threats come from transnational non-state actors such as cyber-criminals and terrorist organisations as well as actor-less threats such as pandemics and climate change.[143] Given the flexibility inherent in Article 51, combined with the centrality of climate change to Western security doctrines, arguments could be explored to test whether Article 51 could be used to justify military action in response to, for instance, an upstream state deliberately blocking shared transboundary water supply critical to sustain life in a downstream state. Indeed, ongoing tensions between Egypt and Ethiopia over the construction of the Grand Ethiopian Renaissance Dam on the Blue Nile help ground such hypothetical explorations in some semblance of reality.[144]

Regional Organisations Acting under Chapter VIII

Chapter VIII of the UN Charter provides the constitutional basis for regional and subregional organisations using military force to maintain international peace and security. However, regional organisations and regional agencies can only undertake enforcement action with the eventual express authorisation of the Security Council, as Article 53 makes clear (authorisation does not need to be given in advance of action). Therefore, as with Chapter VII, Chapter VIII also upholds the pre-eminence of the Council regarding the use of military force.[145] That said, there is a general willingness among the international community that regional organisations can and should deal with human protection catastrophes within their own region. However, there is a general unwillingness to accept regional organisations acting out-of-area against a non-member.[146]

The risks posed by climate change to regional peace and security have been largely accepted by major regional organisations, especially those with responsibilities in climate hotspots such as North Africa. For example, in March 2021 the Peace and Security Council of the African Union (AU) formally acknowledged 'the negative effects of climate change, in general, and in particular, the growing threat they pose to socio-economic development, as well as on sustainable peace, security and stability in Africa, the livelihoods and general wellbeing of the African people'.[147] This was the AU's first communique dedicated to climate security issues.[148] The AU has since proposed a variety of governance solutions to address the climate-conflict nexus.[149] Given the considerable threat posed by climate change to life and livelihoods in North Africa, there may be a need for external support.

The North Atlantic Treaty Organisation (NATO), the world's largest military alliance, has long viewed climate change as a threat multiplier and is currently in the process of building its capability to allow it to function in regions that will be heavily impacted by climate change.[150] For example, in March 2021, NATO's Secretary-General, Jens Stoltenberg, explained that 'climate change will lead to more extreme weather, to droughts and to flooding, force people to move, to more fierce competition about scarce resources, water, land', and therefore, NATO collectively and as individual states need to start reshaping its military to ensure it remains preeminent should deployment in climate impacted zones be required.[151] Air power has been a particular focus. NATO believes that climate change will negatively impact on its air operations in several ways including aircraft performance (e.g. increased temperatures limit engine performance), visual limitations (e.g. increased likelihood of large-scale dust storms in arid regions), and on its airbases (e.g. staff operating conditions during heatwaves).[152] Climate-proofing NATOs aerial assets is an important step should the (likely) need arise to support other regional organisation in climate-related civilian protection operations such as those that provide airborne relief through fixed-wing air drops of food and medical supplies – which would require bearing in mind Chapter VIII authorisation.[153]

General Assembly Authority

The UN General Assembly – unlike the Security Council – has no formal legal authority under the UN Charter to authorise the use of force. However, it can give important political cover for any intervention supported by a broad international consensus. In addition, the General Assembly can play a role in authorising military action in the exercise of its Uniting for Peace procedures (Resolution 377 A (V)) adopted in 1950 to address situations whereby 'the Security Council, because of lack of unanimity of the permanent members, fails to exercise its primary responsibility for the maintenance of international peace and security'.[154] This involves a mechanism whereby 'the General Assembly [if not in session at the time] may meet in a special emergency session' to consider response options upon a request either by a procedural vote in the Security Council (which cannot be blocked by the permanent members) or by a majority of the UN General Assembly members.[155] The Uniting for Peace resolution has been implemented 13 times since 1951 (invoked by the Security Council eight times and the General Assembly five times).

A situation relevant to the climate conversation occurred in December 1971. In this situation, the USSR vetoed draft resolutions S/10416 of 4 December and S/10423 of 5 December blocking the Security Council acting in response to the large-scale massacres occurring against the Bengali population in East Pakistan (present-day Bangladesh), which caused an estimated 8 to 10 million refugees to flee to India. On 6 December, the Security Council passed Resolution 303 referring the question to the General Assembly as provided for in Resolution 377 A (V) Uniting for Peace. As the Twenty-Sixth Regular Session of the General Assembly was in session no 'emergency special session' was necessary and the issue was dealt with under the agenda item UN Assistance to East Pakistan Refugees. The subsequent General Assembly resolution 2790 (XXVI) endorsed the United Nations High Commissioner for Refugees (UNHCR)'s role as the 'focal point' for the UN's East Pakistani Relief Operation.[156] This was the first time in a humanitarian crisis that the UNHCR was chosen to act as general coordinator for all UN assistance. In 2012, the government of Bangladesh honoured the UN refugee agency for providing 'life-saving aid to some 10 million refugees during the 1971 war of independence from Pakistan'.[157]

This example provides a precedent to use the Uniting for Peace procedures to protect refugees fleeing a human protection catastrophe. Present-day Bangladesh is one of the world's most vulnerable countries to climate change impacts. The country of 166 million sits on the floodplain of several major rivers and is also susceptible to seasonal cyclones, storm surges, extreme rainfall, floods, growing salinisation of the soil, and sea level rise.[158] The World Bank estimates that Bangladesh may generate between 3.6 million and 13.3 million 'climate migrants' (people internally displaced because of climate change) by 2050 – the differential based on emissions/temperature and development scenarios – which has clear implications for refugee flows into neighbouring countries.[159] Indications suggest that should Bangladesh 'get hit' by a devastating climate emergency, the Security Council would likely respond.[160] But this is by no means a guarantee of action in Bangladesh or in any other country. Western governments could consider using the Uniting for Peace resolution in response to climate-related human protection calamities should the Security Council fail to act.

Notes

1 UN Secretary-General António Guterres, 'In Flood-Hit Pakistan, Guterres Appeals for "Massive" Global Support, Tougher Action on Climate Change', *United Nations*, 9 September 2022.

2 Ibid.

3 Gareth Evans, *The Responsibility to Protect: Ending Mass Atrocity Crimes Once and For All* (Washington, DC: Brookings Institution Press, 2008), p 105.

4 See for example, 'U.S. Disaster Relief at Home and Abroad', *Council on Foreign Relations*, 11 October 2022. 'Centre for Disaster Protection' (created in 2017 with UKAid funding), 2022; UK-based 'Start Network' linking NGOs for disaster relief. 'Emergency Response Coordination Centre', *European Commission*, 2022. 'New G7 Action to Keep People Safer from Climate Disasters', Press Release, *UK Government*, 13 June 2021. 'Central Emergency Response Fund', *United Nations*, 2022; 'Secretary-General's Video Message on Flash Appeal in Support of Pakistan Flood Response Plan for Pakistan', *United Nations*, 30 August 2022.

5 Evans, *The Responsibility to Protect* (Brookings, 2008), p 105. See also 'World Summit Outcome', paragraph 139.

6 Evans, *The Responsibility to Protect* (Brookings, 2008), p 106.

7 Ibid., p 106–110.

8 Ibid., p 106.

9 'DPPA Around the World', *DPPA* (accessed 2022).

10 Christopher Hodder, 'Climate Change and Security in the United Nations Assistance Mission to Somalia', *Climate Diplomacy*, 26 August 2021. See also, 'Current Presences', *DPPA* (accessed 2022); and 'Addressing the Impact of Climate Change on Peace and Security', *DPPA* (accessed 2022).

11 'Debate on Climate Change, Peace and Security in Africa', *African Union*, 21 April 2022.

12 'Climate Change and Conflict – Three Things You Need to Know', *Crisis Management Initiative*, 16 September 2019.

13 'Jimmy Carter: How to Repair the US–China Relationship and Prevent a Modern Cold War', *Washington Post*, 31 December 2018. See letter by Jimmy Carter in Former US President Emphasises US–China Engagement, *CGTN*, 08 August 2020.

14 The Centre for Humanitarian Dialogue: 'Negotiating Natural Resources for Peace: Ownership, Control and Wealth-Sharing', 2009; 'Reconnecting Mediation: Connecting Tracks, Processes and People', 2019; 'Agro-Pastoral Mediation in the Sahel', 21 September 2021; 'Mediation to Mitigate the Human Cost of Climate Change – Michael Vatikiotis', 8 June 2022.

15 See 'Climate Change', *The Elders* (accessed 2022). See chapters 2 and 3 for remarks by former UN Secretary-General Ban Ki-moon, former Prime Minister of Norway Gro Harland Brundtland, and former President of Ireland, Mary Robinson. See also for example, members of The Elders: The Most Reverend Desmond Tutu, 'We Fought Apartheid. Now Climate Change Is Our Global Enemy', *The Guardian*, 21 September 2014; former First Lady of South Africa Graca Machel, 'Why Ignoring Africa's Climate Change Problem Could Have Devastating Effects', Opinion Editorial, *Time*, 12 September 2019; former UN High Commissioner for Human Rights Zeid Ra'ad Al Hussein, 'Zeid Urges Climate Change Ambition as Paris Deal Enters into Force', Press Release, *OHCHR*, 3 November 2016; Ellin Stein, 'The President [Jimmy Carter] Who Wanted Us to Stop Climate Change', *Slate*, 21 September 2019; former President of Brazil Fernando Henrique Cardoso, 'No Leniency for Deforestation', *Pesquisa*, November 2008; (Felipe Calderón Hinojosa) and former President of Chile Ricardo Lagos, 'Latin American Economies Should Show Leadership on Climate Negotiations', Commentary, *The Economist*, 27 November 2015; former President of Colombia Juan Manuel Santos, 'We Do Not Have to Choose Between a Healthy Economy and a Healthy Planet', *The Elders*, 15 September 2020; former President of Mexico Ernesto Zedillo (Ed.), *Global Warming: Looking Beyond Kyoto* (Washington, DC: Brookings Institute Press, 2008).

16 Evans, *The Responsibility to Protect* (Brookings, 2008), p 108 and 109.

17 'Climate Inaction: An Unconscionable Betrayal of Future Generations', *The Elders*, 7 June 2017.

18 Evans, *The Responsibility to Protect* (Brookings, 2008), p 111.

19 Joshua Keating, 'How Does the U.S. Decide Which Governments to Recognize?', *Foreign Policy*, 13 April 2010.

20 'Syria Signs Agreement with UAE Firms to Build 300-Megawatt Power Station', *Alarabiya News*, 11 November 2021.

21 'US Does Not Support Normalising Ties with Syria, Blinken Says', *Aljazeera*, 13 October 2021.

22 'Australia Committed to Empowering an Effective Pacific Islands Region in The Climate Change Negotiations', *Secretariat of the Pacific Regional Environment Programme*, 17 November 2022.

23 'Action for Climate Empowerment', *UNFCCC* (accessed 2022).

24 Ibid.

25 Douglas Booth, 'Hitting Apartheid for Six? The Politics of the South African Sports Boycott', *Journal of Contemporary History*, Vol 38, No 3, 2003, p 477–493.

26 'Fossil of the Day: Archive', *Climate Action Network International* (accessed 2022).

27 'President Obama Inspires Students with Landmark Speech', *University of Queensland*, 16 November 2014. 'Abbott Government Announces Plan to Cut Emissions by 26 to 28 per cent by 2030', *The Sydney Morning Herald*, 12 August

2015. NB: US pledged to reach this reduction in 2025, compared to Australia in 2030.

28 Evans, *The Responsibility to Protect* (Brookings, 2008), p 114.

29 Ibid., p 114–115.

30 The issue of state neglect discussed in Ashley McLachlan-Bent and John Langmore, 'A Crime against Humanity?: Implications and Prospects of the Responsibility to Protect in the Wake of Cyclone Nargis', *Global Responsibility to Protect,* No 3, 2011, p 40.

31 John Kemp, 'Developing Nations Need Help to Quit Coal Addiction', Opinion Editorial, *Reuters*, 21 July 2021.

32 See for example, Jonathan Masters, 'What Are Economic Sanctions?', *Council on Foreign Relations*, 12 August 2019.

33 Evans, *The Responsibility to Protect* (Brookings, 2008), p 113–114.

34 See for example, Masters, 'What Are Economic Sanctions?'.

35 Evans, *The Responsibility to Protect* (Brookings, 2008), p 113–114.

36 See for example, Joy Gordon, 'The Not So Targeted Instrument of Asset Freezes', *Ethics and International Affairs*, Vol 33, No 3, 2019, p 304.

37 Phillip Brown, 'Oil Market Effects from U.S. Economic Sanctions: Iran, Russia, Venezuela', *Congressional Research Service*, 5 February 2020, p 4–8.

38 Gordon, 'The Not So Targeted Instrument of Asset Freezes', p 306.

39 Brown, 'Oil Market Effects from U.S. Economic Sanctions', p 10–18.

40 Gordon, 'The Not So Targeted Instrument of Asset Freezes', p 306.

41 Masters, 'What Are Economic Sanctions?'.

42 Search countries on 'European Union Sanctions', *European Union External Action* (accessed 2022).

43 'EU Iran Sanctions: Ministers Adopt Iran oil Imports Ban', *BBC*, 23 January 2012' 'EU Follows US Ban on Oil Imports, and Expands List of Designated Persons', *HFW (legal services)*, September 2011. See also, 'EU Response to the Syrian Crisis', *European Union External Action*, 26 April 2019.

44 'Sanctions FAQ', *Norton Rose Fulbright*, March 2015.

45 Jan Rocha, 'Europe Warns of Brazilian Trade Boycott Over', *Climate News Network*, 21 September 2020.

46 Evans, *The Responsibility to Protect* (Brookings, 2008), p 115.

47 Ibid.

48 'Global Climate Litigation Report: 2020 Status Review', *UN Environment Programme*, 26 January 2021.

49 'Ibid., p 13.

50 'Urgenda Foundation v. State of the Netherlands', *Climate Change Litigation Databases*, 2022. See also, 'Global Climate Litigation Report: 2020 Status Review', *UN Environment Programme*, 26 January 2021, p 43.

51 Noni Shannon and Elisa de Wit, 'Urgenda Foundation v Netherlands: Historic Climate Change Decision Upheld', *Norton Rose Fulbright*, December 2019.

52 'Global Climate Litigation Report: 2020 Status Review', *UN Environment Programme*, 26 January 2021, p. 15–17. See also 'Australian Court Finds Government Has Duty to Protect Young People from Climate Crisis', *The Guardian*, 27 May 2021; and 'Sussan Ley Does Not Have Duty of Care to Protect Young from Climate Crisis, Appeal Court Rules', *The Guardian*, 15 March 2022.

53 'Global Climate Litigation Report: 2020 Status Review', UN Environment Programme, 26 January 2021, p 15.

54 Ibid., p 31.
55 'In re Court on Its Own Motion v. State of Himachal Pradesh and Others', *Climate Change Litigation Database*, 2022.
56 Ibid. See also, 'Environmental Courts and Tribunals 2021: A Guide for Policymakers', *UN Environment Programme*, 2022, p 28.
57 George Pring and Catherine Pring, 'Environmental Courts and Tribunals: A Guide for Policy Makers', Report, *UN Environment Programme*, 2016, p 1.
58 Linda Yanti Sulistiawati, Farah Bouquelle, Jolene Lin, Luc Lavrysen, Mark Ortega, Ricardo Pereira, and Sean Tseng, 'Environmental Courts and Tribunals 2021: A Guide for Policymakers', *UN Environment Programme*, 2022, p i.
59 Pring and Pring, 'Environmental Courts and Tribunals', p iv.
60 Daniel Billy and others v Australia (Torres Strait Islanders Petition), *Climate Change Litigation Databases*, 2020.
61 'Global Climate Litigation Report', p 14.
62 Ibid., p 14–15.
63 Evans, *The Responsibility to Protect* (Brookings, 2008), p 120. See also former UN Under-Secretary of Peacekeeping Operations, Marrack Goulding, *Peacemonger* (Baltimore: John Hopkins University Press, 2003), p 14–15.
64 See 'Chapter VI: Pacific Settlement of Disputes', *United Nations Charter*.
65 Evans, *The Responsibility to Protect* (Brookings, 2008), p 120–121.
66 Ibid., p 123. See also, 'The Report of the Panel on United Nations Peace Operations ("The Brahimi Report")', A/55/305–S/2000/809, *United Nations*, 21 August 2000.
67 Ibid., p 124.
68 Aimée-Noël Mbiyozo and Ottilia Anna Maunganidze, 'Climate change and Violence in Africa: No Time to Lose', *Institute for Security Studies*, 17 March 2021.
69 Ibid.
70 Florian Krampe, 'Why United Nations Peace Operations Cannot Ignore Climate Change', *Stockholm International Peace Research Institute*, 22 February 2021.
71 Ibid.
72 See for example, Part 1 in 'Greening the Blue Helmets: Environment, Natural Resources and UN Peacekeeping Operations', *UN Environment Program*, May 2012. See also, Lucile Maertens, 'From Blue to Green? Environmentalization and Securitization in UN Peacekeeping Practices', *International Peacekeeping*, Vol 26, No 3, p 302–326.
73 'Preventing, Mitigating & Resolving Transhumance-Related Conflicts in UN Peacekeeping Settings: A Survey of Practice', *UN Department of Peace Operations*, 2020.
74 Agathe Sarfati, 'Toward an Environmental and Climate-Sensitive Approach to Protection in UN Peacekeeping Operations', *International Peace Institute*, October 2022, p 7.
75 'Preventing, Mitigating & Resolving Transhumance-Related Conflicts in UN Peacekeeping Settings: A Survey of Practice', *UN Department of Peace Operations*, 2020.
76 Part 2 'Greening the Blue Helmets: Environment, Natural Resources and UN Peacekeeping Operations', *UN Environment Program*, May 2012.
77 Sarfati, 'Toward an Environmental and Climate-Sensitive Approach to Protection in UN Peacekeeping Operations', p 2.
78 Ibid.

79 Paul R. Williams, Jonathan Worboys, and J. Trevor Ulbrick, 'Preventing Mass Atrocity Crimes: The Responsibility to Protect and the Syria Crisis', *Case Western Reserve Journal of International Law*, Vol 45, No 1, 2012, p 449. See also, Evans, *The Responsibility to Protect* (Brookings, 2008), p 125; Anna Louise Strachan, 'Safe Haven', Humanitarianism: Keywords, 2020; and Phil Orchard, 'Revisiting Humanitarian Safe Areas for Civilian Protection', *Global Governance*, 20, 2014, p 55.
80 Evans, *The Responsibility to Protect*, p 125.
81 Rutger Birnie and Jennifer Welsh, 'Displacement, Protection and Responsibility: A Case for Safe Areas', *Global Responsibility to Protect*, 10, 2018, p 338.
82 Evans, *The Responsibility to Protect* (Brookings, 2008), p 126. See also, Daniel Jacob, 'Safe Areas and the Responsibility to Protect', *Global Responsibility to Protect*, Vol 10, 2018, p 314.
83 Strachan, 'Safe Haven'.
84 'Virunga Alliance', Virunga National Park (accessed 2022). See also, Judith Verweijen and Saidi Kubuya Batundi, 'Virunga National Park: Rethinking 'Law and Order' In Conservation', *London School of Economics*, 5 July 2019; 'More Safe Havens for Threatened Species', *Joint Media Release*, Australian Government, 16 November 2020.
85 'No Redd in Africa Network: Africans Unite Against New Form of Colonialism', *African Conservation Foundation*, 3 April 2013.
86 Sara Peach, 'Where Should I Move to Be Safe from Climate Change?', *Yale Climate Connections*, 20 December 2019.
87 Radhika Murti and Camille Buyck (Eds.), 'Safe Havens: Protected Areas for Disaster Risk Reduction and Climate Change Adaptation', *International Union for Conservation of Nature*, 2014.
88 Paul R. Williams, Jonathan Worboys, and J. Trevor Ulbrick, 'Preventing Mass Atrocity Crimes: The Responsibility to Protect and the Syria Crisis', *Case Western Reserve of International Law*, Vol 45, No 1, 2012, p 499–500; Orchard, 'Revisiting Humanitarian Safe Areas for Civilian Protection', p 64 and 61. Birnie and Welsh, 'Displacement, Protection and Responsibility', p 334, 350, and 352–353.
89 'Climate Change and Disaster Displacement', *UN High Commissioner for Refugees*, 21 August 2017. See also 'Climate Change and Disaster Displacement in the Global Compact on Refugees [adopted by the UNGA on 17 December 2018]', UN High Commissioner for Refugees.
90 For a variety of forecasts see Baher Kamal 'Climate Migrants Might Reach One Billion by 2050', *Interpress Service*, 2017. See also, 'Climate Crisis Could Displace 1.2bn People by 2050, Report Warns', *The Guardian*, 9 September 2020.
91 Karl P. Mueller, 'Denying Flight Strategic Options for Employing No-Fly Zones', *RAND Corp*, 2013, p 2. See also, Evans, *The Responsibility to Protect* (Brookings, 2008), p 126.
92 Mueller, 'Denying Flight Strategic Options for Employing No-Fly Zones', 2013, p 6.
93 Evans, *The Responsibility to Protect* (Brookings, 2008), p 126.
94 'Typhoon Haiyan: US Carrier Boosts Philippines Relief Effort', *BBC*, 14 November 2013. Ben Parr and Don Henry, 'Philippines Climate Change Agenda: High Vulnerability! High Ambition?', *University of Melbourne*, 1 March 2016.
95 'Against All Odds', *World Food Program* (accessed 2022). See also US-based organisation 'Operation Airdrop' (accessed 2022).
96 'Nobody Wants to Drop Food from a Plane. But It's Happening', *NPR*, 24 March 2017.

97 Jeannie L Sowers, Erika Weinthal, and Neda Zawahri, 'Targeting Environmental Infrastructures, International Law, and Civilians in the New Middle Eastern Wars', *Security Dialogue*, Vol 48, No 5, 2017, p 416–424. See also, '2016 Humanitarian Needs Overview – Yemen', *OCHA*, 22 November 2015.

98 'Darfur in Flames, Atrocities in Western Sudan: Abuses in Darfur by Government Forces', *Human Rights Watch*; and SHRO/COHRE v. Sudan (AComHPR), *Weapons Law Encyclopedia*, 2022. See also, 'Sudan Aircraft Bomb Settlement Near Chad – Rebels', *Reuters*, 3 June 2009.

99 'Investigation by the Office of the United Nations High Commissioner for Human Rights on Libya: Detailed Findings', *Human Rights Council*, 23 February 2016; 'Investigation by the Office of the United Nations High Commissioner for Human Rights on Libya', *Human Rights Council*, 15 February 2016.

100 'Convention on the Prohibition of Military or Any Other Hostile Use of Environmental Modification Techniques (ENMOD)', *UN Office for Disarmament Affairs*, 2022.

101 Sowers, Weinthal, and Zawahri, 'Targeting Environmental Infrastructures', p 413–414.

102 Evans, *The Responsibility to Protect* (Brookings, 2008), p 126–127.

103 'Social Media in Peace Mediation a Practical Framework', *UN DPPA (MSU)* with *Swiss Peace*, June 2021, p 2. See also, 'Digital Technologies and Mediation in Armed Conflict', *UN DPPA (MSU)* with *The Centre for Humanitarian Dialogue*, March 2019.

104 Ibid.

105 Ibid.

106 'Facebook Admits It Was Used to Incite Violence in Myanmar', *The New York Times*, 6 November 2018. See also, 'Myanmar's Military Struggles to Control the Virtual Battlefield', Report No 314, *International Crisis Group*, 18 May 2021.

107 Maude Morrison and Adam Cooper, 'In War Zones, Social Media Disinformation Is Costing Lives', *Wired UK*, 20 April 2021.

108 Pip Cook, 'Why Peace Talks Need "Digital Ceasefires"', *Geneva Solutions*, 14 July 2021.

109 'Myanmar Cyclone Nargis Victims Re-connected via Telecommunication', *Reliefweb*, 16 May 2008.

110 '"I Want to Help My Own People": State Control and Civil Society in Burma after Cyclone Nargis', *Human Rights Watch*, 2010, p 29–30.

111 See Emma Larkin, *Everything Is Broken: A Tale of Catastrophe in Burma* (New York: The Penguin Press, 2010).

112 'Burmese journalists Continue to be Arrested, Foreign Journalists still Unwelcome', *Reporters Without Borders*, 25 June 2008.

113 '"I Want to Help My Own People": State Control and Civil Society in Burma after Cyclone Nargis', *Human Rights Watch*, 2010, p 29–30. See also, 'Call for Government to Allow Foreign Journalists to Cover Disaster', Media Alert, *Committee to Protect Journalists*, 7 May 2008.

114 'Myanmar Cyclone Nargis Victims Re-connected via Telecommunication', *Reliefweb*, 16 May 2008.

115 Evans, *The Responsibility to Protect* (Brookings, 2008), p 128.

116 Ibid.

117 Ibid., p 139–147.

118 Ibid., p 141. See also, *The Responsibility to Protect*, ICISS p 35–37; and *A More Secure World*, High-Level Panel, p 66–67, 106–107.

119 Robyn Eckersley, 'Ecological Intervention: Prospects and Limits', *Ethics and International Affairs*, Vol 21, No 3, p 293–316. See also, Linda A. Malone, 'Green Helmets: Eco-intervention in the Twenty-First Century', *Faculty Publications,* William and Mary Law School, 2009.

120 Ibid.

121 Ibid.

122 'Amazon Rainforest Fires: Global Leaders Urged to Divert Brazil From "Suicide" Path', *The Guardian*, 23 August 2019.

123 'Amazon Rainforest Now Emitting More CO_2 Than It Absorbs', *The Guardian*, 15 July 2021; 'Rainforest Mafias: How Violence and Impunity Fuel Deforestation in Brazil's Amazon', *Human Rights Watch*, 17 September 2019.

124 'France's Macron Says Real "Ecocide" Going on in Amazon', *Reuters*, 24 August 2019.

125 'G7 Leaders Agree Plan to Help Amazon Countries Fight Wildfires', *The Guardian*, 27 August 2019.

126 'Brazil's Bolsonaro Accuses French President of "Colonialist Mindset" After Calls for Action on Amazon Fires', *CNBC*, 23 August 2019. See also, 'Brazil Rejected a $20 Million Donation from the G7 to Fight the Amazon Fires, Calling the Gesture "Colonialist and Imperialist"', *Insider*, 27 August 2019.

127 Stephen M. Walt, 'Who Will Save the Amazon (and How)?', *Foreign Policy*, 5 August 2019.

128 Lawrence Douglas, 'Do the Brazil Amazon Fires Justify Environmental Interventionism?', *The Guardian*, 31 August 2019. See also, Gustavo Macedo, 'The Responsibility to Protect the Amazon', Opinion, *E-International Relations*, 31 March 2021.

129 'United Nations Charter, Chapter VII: Action with Respect to Threats to the Peace, Breaches of the Peace, and Acts of Aggression', Article 39, 41 and 42, United Nations.

130 Evans, *The Responsibility to Protect* (Brookings, 2008), p 133.

131 Ibid., p 134.

132 See for example, UK President of the UNSC opening address at 5663rd meeting, Tuesday, 17 April 2007, 10 a.m. New York; and, Peter Wittig, Permanent Representative Germany, 'Concept Note', Letter dated 1 July 2011; and Ban Ki-moon, Remarks to the Security Council on the Impact of Climate Change on International Peace and Security, UN Secretary-General, 20 July 2011. More recently, see Virtual Security Council Debate on Climate and Security, 23 February 2021: British Prime Minister, Boris Johnson: 'It is absolutely clear that climate change is a threat to our collective security and the security of our nations'. Further, if the suffering and strife caused by climate change was done at the hand of 'some kind of despotic warlord or civil war, then nobody would question the right and the duty of this U.N. Security Council to act'. US Climate Envoy, John Kerry: 'the climate crisis is indisputably a Security Council issue … The climate threat is so massive, so multifaceted … we bury our heads in the sand at our own peril'. French President, Emmanuel Macron: 'The fight against climate change and for the protection of the environment is a matter of peace and security … I call on the Security Council to take on these issues as part of its mandate'. China's Climate Envoy, Xie Zhenhua: 'Climate change has become a pressing and serious threat to the survival, development and security of humankind'. Russia's Representative to the U.N., Vasily Nebenzya: the link between climate change and conflicts

was specific to certain countries and there was 'no justification' for making that connection globally. See also Christopher K. Penny, 'Climate Change as a "Threat to International Peace and Security"', in Shirley V. Scott and Charlotte Ku (Eds.), *Climate Change and the UN Security Council* (Edward Elgar, 2017).

133 'Climate Security at the UNSC – A Short History', *Climate Diplomacy* and *Adelphi* (accessed 2022).

134 Florian Krampe and Cedric de Coning, 'Russia's "Nyet" Does Not Mean Climate Security Is Off the Security Council Agenda', *SIPRI*, 13 December 2021; 'How UN Member States Divided Over Climate Security', *International Crisis Group*, 22 December 2021; 'Russia Blocks U.N. Move to Treat Climate as Security Threat', *The New York Times*, 13 December 2021. See also on the issue of climate-conflict contestation in relation the UN Security Council, Ken Conca, Joe Thwaites, and Goueun Lee, 'Climate Change and the UN Security Council: Bully Pulpit or Bull in a China Shop?', *Global Environmental Politics*, Vol 17, No 2, May 2017, p 1–20; and Ken Conca, 'Is There a Role for the UN Security Council on Climate Change?', *Environment: Science and Policy for Sustainable Development*, Vol, No 1, 2019, p 4–15.

135 'Remarks by Ambassador Linda Thomas-Greenfield After the Vote on a UN Security Council Resolution on Climate and Security', United States Mission to the United Nations, 13 December 2021.

136 See for example, Wim Zwijnenburg and Brittany Rose, 'Signs of Progress: Environment and Conflict in the UN Security Council', *Planetary Security Initiative*, March 2020.

137 See for example, 'Time for the UN Security Council to Act on Climate Security', Statement, *International Crisis Group*, 7 December 2021.

138 'United Nations Charter, Chapter VII: Action with Respect to Threats to the Peace, Breaches of the Peace, and Acts of Aggression', Article 51, United Nations.

139 Christine Chinkin and Mary Kaldor, *Self-Defence as a Justification for War: The Geo-Political and War on Terror Models* (Cambridge: Cambridge University Press, 2017), p 136.

140 See for example, Daniel Bethlehem, 'Principles Relevant to The Scope of a State's Right of Self-Defense Against an Imminent or Actual Armed Attack by Nonstate Actors', *American Journal of International Law*, Vol 106, 2012, p 770–777.

141 Evans, *The Responsibility to Protect* (Brookings, 2008), p 134. See also Christine Chinkin and Mary Kaldor, *Self-Defence as a Justification for War*, 2017, p 157–161.

142 Ibid.

143 'Interim National Security Strategic Guidance', *The White House*, March 2021, p 7; 'Peace and Security in 2022: Overview of EU Action and Outlook for the Future', *European Parliament*, July 2022, p I–II.

144 See for example, 'Nile Dam Talks: Unlocking a Dangerous Stalemate', Statement, *International Crisis Group*, 16 March 2020.

145 'United Nations Charter, Chapter VIII: Regional Arrangements', Article 53, United Nations. Evans, *The Responsibility to Protect* (Brookings, 2008), p 135.

146 Evans, *The Responsibility to Protect*, p 135.

147 Peace and Security Council of the African Union, 984th Meeting at the Level of Heads of State and Government on 'Sustainable Peace in Africa: Climate Change and Its Effects on Peace and Security in the Continent', PSC/AHG/COMM. 1, Communique, 9 March 2021, p 1.

148 Aimée-Noël Mbiyozo and Ottilia Anna Maunganidze, 'Climate Change and Violence in Africa: No Time to Lose', *Institute for Security Studies*, 17 May 2021.
149 'African Union Climate Change and Resilient Development Strategy and Action Plan', *African Union*, 2022, p 36–37.
150 Rene Heise, 'NATO Is Responding to New Challenges Posed by Climate Change', *NATO*, 1 April 2021.
151 Remark by NATO Secretary-General Jens Stoltenberg 'Facing Sweltering Soldiers and Flooded Ports, NATO to Focus on Climate Change', *The Washington Post*, 24 March 2021.
152 Heise, 'NATO Is Responding to New Challenges Posed by Climate Change'.
153 NATO's most prominent airborne mission guided by Chapter VIII occurred in March 2011 with the cooperative endeavour between NATO and the Arab League to implement a no-fly zone over Libya to help protect civilians and maintain international peace and security. See for example, 'Security Council Approves 'No-Fly Zone' Over Libya, Authorizing 'All Necessary Measures' to Protect Civilians, by Vote of 10 in Favour with 5 Abstentions', SC/10200, *Security Council*, 17 March 2011; 'NATO No-Fly Zone Over Libya, Operation Unified Protector', Fact Sheet, *NATO*, 2011.
154 Resolution 377 A (V), United Nations General Assembly, adopted 3 November 1950. See also, Evans, *The Responsibility to Protect* (Brookings, 2008), p 136.
155 Evans, *The Responsibility to Protect*, p 136
156 United Nations assistance to East Pakistan refugees through the United Nations focal point and United Nations humanitarian assistance to East Pakistan 2790 (XXVI), 06 December 1971. See also, Phil Orchard, *Protecting the Internally Displaced: Rhetoric and Reality* (Oxon: Routledge, 2019), p 90–92.
157 'UNHCR Honoured by Bangladesh for Helping Millions in 1971 Conflict', *UNHCR*, 27 March 2012.
158 'Intended Nationally Determined Contributions (INDC)', Ministry of Environment and Forests (MOEF) Government of the People's Republic of Bangladesh (lodged to the UNFCC), September 2015, p 9–10. Amelie Bernzen, J. Craig Jenkins, and Boris Braun, 'Climate Change-Induced Migration in Coastal Bangladesh? A Critical Assessment of Migration Drivers in Rural Households under Economic and Environmental Stress', *Geosciences*, Vol 9, No 51, 2019, p 1–20.
159 Kanta Kumari Rigaud, Alex de Sherbinin, Bryan Jones, Jonas Bergmann, Viviane Clement, Kayly Ober, Jacob Schewe, Susana Adamo, Brent McCusker, Silke Heuser, and Amelia Midgley, 'Groundswell: Preparing for Internal Climate Migration', *World Bank*, Vol 1, 2018, p 148.
160 Interviews with relevant officials in Ken Conca, Joe Thwaites, Goueun Lee, 'Climate Change and the UN Security Council: Bully Pulpit or Bull in a China Shop?', *Global Environmental Politics*, Vol 17, No 2, May 2017, p 13. See also, Ken Conca, 'Is There a Role for the UN Security Council on Climate Change?', *Environment: Science and Policy for Sustainable Development*, Vol 61, No 1, 2019, p 12.

6 Climate Action and the Responsibility to Rebuild

> The responsibility to rebuild [is] to provide, particularly after a military intervention, full assistance with recovery, reconstruction and reconciliation, addressing the causes of the harm the intervention was designed to halt or avert.
>
> (*Report of the International Commission on Intervention and State Sovereignty*, 2001, p xi)

> As many know, the Chinese expression for 'crisis' consists of two characters side by side. The first is the symbol for 'danger', the second the symbol for 'opportunity'.
>
> (Former US Vice President Al Gore, *An Inconvenient Truth*, 2006, p 3)

The best single indicator of future conflict is past conflict.[1] History repeatedly shows that a sustainable peace cannot be guaranteed simply because a peace agreement has been reached or military victory achieved.[2] The 'responsibility to rebuild' seeks to address the problem of recurring human protection catastrophes by, in a general sense, encouraging decision-makers to take seriously and commit adequate resources, expertise, and time to rebuilding societies in the aftermath of war or atrocities; and in a more specific and practical sense, by identifying a variety of policy options for decision-makers to consider within four interrelated but distinct dimensions: achieving security, good governance, justice and reconciliation, and economic and social development.[3] The responsibility to rebuild phase is considered the beginning of a new process of conflict prevention (see Chapter 4), with again the focus on structural prevention and tackling the long-term root causes of the violence.[4] Chapter 6, the final chapter of this book, explores each dimension in sequence, paralleling and integrating climate action issues and options where possible.

Environmental peacebuilding provides a solid conceptual backbone to discuss climate-informed rebuilding in the aftermath of conflict. While on occasions defined as spanning prevention, response, and recovery, environmental peacebuilding, narrowly defined, focuses on 'conflict-sensitive and sustainable management of renewable natural resources in conflict-affected or

DOI: 10.4324/9780367201364-9

post-conflict states' that support the ecological underpinnings of a sustainable peace.[5] In addition, as a concept, environmental peacebuilding has its roots in post-conflict situations, being developed throughout the 1990s, with a focus on governance solutions to intrastate 'resource wars' in post-conflict contexts.[6] Scholarly attention on this topic in the early 2000s[7] contributed to the UN Environment Programme's first post-conflict environmental assessment in Afghanistan in 2003, which 'provided evidence that the environment could also serve as a critical peacebuilding tool in "post-conflict" reconstruction'.[8] More contemporary compilations of work also tend to focus on this phase of the conflict cycle. For example, many of the cases in the 2021 *International Affairs* Special Issue on environmental peacebuilding gravitate towards post-conflict reconstruction.[9] Some of the key examples and ideas from this body of work are used throughout this chapter.

Post-crisis situations present an opportunity to rebuild societies and economies in a low-pollution and climate-resilient way. For example, consideration should be given to distributing off-grid solar with battery storage rather than rebuilding fossil fuel-based energy systems; and or ensuring that transport infrastructure is rebuilt with an eye to future climate impacts such as storms and sea level rise. Such forward looking strategies give the country's children – who are least responsible for the deadly violence and or the climate impact – a greater chance to survive, and possibility even thrive, as their world warms.[10]

Achieving Security

Peacekeeping in Support of Peacebuilding

Since the Cold War, major 'peacekeeping' and 'peace enforcement' missions have tended to comprise a 'peacebuilding' component – for example, in Haiti, Sierra Leone, and Timor-Leste; and in Bosnia, Afghanistan, and Iraq, respectively.[11] By way of substantive definitions and differences: 'peacekeeping' missions are generally UN-led, but on occasions can be jointly led with regional organisations, and are deployed with the consent of the major parties to the conflict. They are tasked with performing impartial monitoring, supervision, and verification of ceasefire agreements, among other such duties discussed below.[12] By contrast, 'peace enforcement' missions are not premised on the consent of all sides and do not require impartiality. They are expressly authorised under Chapter VII and involve actions, including the use of military force, to maintain or restore peace and security. UN 'blue-helmeted' peace enforcement missions are extremely rare, the most prominent example taking place in Korea in the 1950s. More often a multinational force will be deployed under a UN mandate, for example, the US-led response to the Iraqi invasion of Kuwait in 1991.[13] In-between traditional peacekeeping and peace enforcement missions sits a more robust peacekeeping mission, sometimes described as 'coercive peacekeeping' or 'peacekeeping plus', which allows peacekeepers to respond to violence against civilians when and where they see it.[14] 'Peacemaking', to add a further term discussed below, is defined as 'the use of diplomatic means [such a mediation and dialogue] to persuade hostile parties to settle a dispute which has led, or could lead, to armed conflict'.[15]

'Peacebuilding' is completely different again, and is generally understood as a long-term process that aims to deliver a sustainable peace by tackling the root causes of a potential or past conflict.[16] The root causes of conflict are legion and can span human rights; ethnic or religious discrimination; issues related to justice, law and order; the distribution of wealth and opportunity; and environmental factors including climate change – as 'environmental peacebuilding' emphasises.[17] Peacebuilders will attempt to address the root causes of conflict in collaboration with local and international actors, by renovating or rebuilding the apparatuses of state power such as governance structures and law enforcement; as well as, setting the conditions for economic reconstruction, development, and growth. There are a variety of security tasks that peacekeeping missions must be prepared to perform in support of peacebuilding activities.[18] First, peacekeeping personnel are required to support a peacemaking process towards a peace agreement being reached. Tasks may include providing security for peacemakers and conflicting parties during the negotiations. Second, peacekeepers are generally expected to provide immediate post-agreement security and stability. Tasks may include responding to 'spoilers' seeking to disrupt the peace, sometimes with force if necessary and mandated under a 'peacekeeping plus' initiative. Third, peacekeepers may be required to provide a longer-term security function to support the restoration of good governance, the rule of law, and economic recovery. Tasks may include providing a general policing role to create a safe and stable environment.

Climate INGOs and government agencies have published a raft of practical guides for peacebuilders to use in post-conflict situations. Some contain a list of climate-related questions for peacebuilders to consider, others contain checklists, others present integrated climate-conflict risk assessment models, while others provide scenario planning advice – some contain all the above.[19] None of these guides however appear to discuss the role of peacekeepers. To remedy this, during peace negotiations, peacekeepers could provide security to climate specialists tasked with collecting and providing information to the conflicting parties on important local climate issues such as water resources and arable land – both of which would be crucial provisions in any peace agreement (and a subsequent constitution making process if required). Over the mid-term, peacekeepers could provide ongoing support to environmental risk assessment teams tasked with examining the habitability of zones and regions identified as ready for the return of refugees and internally displaced persons.[20] Over the longer term, peacekeepers could help enforce new and old laws regarding the allocation and use of natural resources, as well as help provide a safe and secure environment that encourages donor and industry investment in low-emission enterprises.

Disarmament, Demobilisation, and Reintegration

Disarmament, Demobilisation, and Reintegration (DDR) of former combatants is another key task performed by international peacekeeping missions, working in close cooperation with national authorities, aimed at interrupting the conflict cycle and facilitating the transition of armed groups back into civilian life.[21] Regarding terminology and aims.[22] Disarmament in a post-conflict peacebuilding situation is aimed at the collection and destruction of existing weapons and the progressive suppression of the capacity to produce or purchase new ones. Demobilisation refers to the process of standing down and disbanding armed forces, which may include retaining or reclassifying some personnel, while ensuring that others are transitioned to opportunities in the community that will generate an income for them and their families. Reintegration involves a sustained effort to develop employment, training, and education schemes and opportunities for ex-combatants in the hope of permanently detaching them from the more extreme and violent elements of their recent past.

Reintegration provides the most scope to graft-on climate action activities.[23] Public works are almost always included in reintegration programmes because post-conflict situations generally require infrastructure to be rebuilt and donors can more easily justify the associated expenditure domestically by pointing to a physical structure such as a building, roadway, or bridge.[24] Transport infrastructure as well as energy infrastructure and industrial facilities should all be restored or rebuilt with climate adaptation and mitigation risks and opportunities in mind. Vocational training is another essential component of any reintegration effort. Such training could be geared towards gradually building a workforce with the skills needed to build, refurbish, and maintain low-emissions infrastructure.[25] Natural resources and their use are also central to reintegration efforts across a globe[26] because they provide emergency employment as well as livelihoods for ex-combatants in agriculture, forestry, fisheries, water, and extractive industries. Integrating climate action considerations into these sectors will help shield them from the impacts of climate change and thus support the sustainability of reintegration efforts.[27] Emphasis could also be placed on ecosystem restoration, protecting environmental assets, and ecotourism.[28]

Security Sector Reform

Security Sector Reform (SSR) is a critical task in most peacebuilding contexts.[29] Armies, intelligence agencies and the police may still suffer from some or all the problems that may have contributed to the initial outbreak of the conflict, for instance, poor leadership, resourcing, and accountability; corruption; and/or ethnic imbalance. In addition, regime officials may have used the security sector to oppress (or worse) parts of the population, which may mean that some groups still see the sector as predatory.[30] Post-conflict environments present a window of opportunity to significantly reform the security sector because SSR may have been built into a peace agreement and broader reform plan; internal parties, weakened by the conflict, may be willing to compromise; and external actors may have been invited to be involved.[31]

As the world warms, security sector actors will be increasingly called upon to help communities prepare for the impacts of climate change and to respond when disaster strikes.[32] To perform effectively, security actors will require new climate-informed training programmes, budget streams and oversight processes, as well as conceptual tools and practical equipment. Western governments could consider integrating climate action considerations into their existing SSR programmes. Specific opportunities could include providing security actors with training and access to up-to-date Disaster Risk Reduction tools such as scenario planning methodologies, weather forecasting equipment, and early warning and response systems, and the provision of enhanced search and rescue equipment, training, and processes.[33] Over the longer-term, a strong and fully functional security sector free from corruption can have a significant stabilising effect on the country, which is essential to attract foreign investment in low-emissions and climate-resilient activities.

Achieving Good Governance

Rebuilding Institutions of Governance

Rebuilding the institutions of governance in post-conflict environments will generally require, first, the immediate restoration of government services, and second, the creation or recreation of a system of governance with functioning executive, legislative, and judicial institutions.[34] Restoring public services should consider the opportunities and risks posed by climate change. For instance, it may be quicker and more cost-effective (as well as environmentally responsible), to roll-out decentralised off-grid solar capacity with batteries rather than rebuild destroyed centralised coal-fired power plants and reconnect the associated transmission lines. Rebuilding roads and other transportation networks should also consider low-emission options as well as the risks of climate change such as coastal inundation, sea level rise, and extreme weather events. Climate-sensitive water collection and distribution networks could also be prioritised, replacing inefficient systems. The structural ingredients for good governance include fair constitutional structures, effective human rights protections, the rule of law, and the eradication of corruption – all of which should seek to integrate the climate action agenda (see Chapter 4). Many Western countries are currently implementing a 'climate mainstreaming' agenda, which establishes that the decisions made by the government and their departments and agencies must consider the risks and opportunities posed by climate change. Western donors could prioritise climate mainstreaming (also known as 'good climate governance') in post-conflict peacebuilding contexts.

Climate considerations should also be visible when providing support to help develop an independent civil society (NGOs, business community and academic institutions) capable of actively and safely engaging with government policymaking processes.

Maximising Local Ownership

Any multilateral peacebuilding mission must hand over authority to internal actors and depart the field as soon as functionally possible. Doing so will help avoid a build-up of neo-colonial resentment, inertia, and resistance in the local population.[35] Ultimately, the exercise of authority, and the responsibility taken for those decisions, is for the local population to contend with. However, not overstaying one's welcome should be balanced with the overarching need and purpose of the rebuilding mission in the first place: to embed a variety of properly functioning, effective, and durable civilian protection programmes in the society and economy.[36] Similarly, it would seem appropriate that Western peacebuilding programmes that include a climate action dimension should be guided by the principle of 'maximising local ownership' while also ensuring that oversight, safeguards, and assurances (i.e. for funding) are not entirely removed. Empowering domestic actors at a national, subnational, and community level (where appropriate – for instance, in Syria[37] and Myanmar,[38] it may be preferable to support local communities rather than national governments) is more likely to yield the sustainable management of a climate action programme. However, Western governments are rightly anxious about providing ongoing climate finance and technology without financial controls and transparency and assurances of positive environmental outcomes.[39] One approach that allows donor oversight and evaluation while building local ownership over time is to focus on the creation of long-term climate action plans, which specifies that ongoing access to funding and support is premised on keeping to a *process,* rather than a predetermined outcome.[40]

Achieving Justice and Reconciliation

To achieve justice and reconciliation in the aftermath of a mass atrocity event, peacebuilding missions should consider accomplishing four key tasks.[41] First, rebuilding the justice system, which involves restoring an effective court and correctional system staffed with well-trained lawyers, judges, legal professionals, and police. Second, managing transitional justice, which can involve the immediate prosecution of perpetrators or postponement to allow for national healing, symbolic or substantive reparations, or establishing a truth commission. Third, negotiating the role that traditional justice mechanisms will play such as informal tribe, local, grassroots and/or village level justice systems. Fourth, managing refugee returns – which of the four, presents the most obvious entry point for climate action advocates.

Managing Refugee and IDP Returns

In the aftermath of a mass atrocity situation (or a major climate-related disaster), officials on the ground will likely be confronted with the task of managing the return of refugees and internally displaced persons (IDPs). Therefore, one of the highest of all priorities in post-conflict situations must be to put in place the conditions under which refugees and IDPs can safely and sustainably return to their homes.[42] In 2020, climate-related disasters, especially intense storms and floods, triggered 30.7 million new IDPs, with violent conflict producing 10 million more.[43] By 2030, the number of people displaced by climate events globally is expected to be between 200 million and 1 billion people.[44] Even under conservative estimates, peacebuilders will face a series of new challenges when dealing with mass people movement and their sustainable repatriation. For example, peacebuilders will increasingly need to undertake an assessment process that considers the ongoing habitability of an area.[45] In 2020, the United Nations High Commissioner for Refugees (UNHCR) and Postdam Institute for Climate Impact Research, a leading climate impact assessor, investigated the relationship between climate change and refugee returns. They found that 'severe climate impacts threaten the cornerstones of durable solutions for refugees and internally displaced persons by rendering areas of return and of local integration too dangerous to live in or too fragile to support large populations'.[46] Western governments could continue to develop and help implement climate-informed responses to displaced persons in collaboration with local communities and international organisations.[47]

Achieving Economic and Social Development

Economic Recovery and Development

Economic recovery and development in post-conflict situations require peacebuilders and their international partners to accomplish two substantial economic tasks: first, the immediate economic stabilisation of the country in question, and second, and the much bigger task, to move from basic stabilisation to creating the conditions for sustained economic growth.[48] To have a realistic chance of achieving these two very large tasks, peacebuilders should aim to achieve several smaller-scale ancillary tasks – each of which can include a climate action dimension.

To achieve immediate economic stabilisation, the following two ancillary tasks are important.[49] First, stabilise the currency, which can be achieved with technical support and emergency finance available through the International Monetary Fund (IMF). The IMF's Rapid Financing Instrument (RFI) – which recently replaced the IMF's Emergency Natural Disaster Assistance and Emergency Post-Conflict Assistance programme – provides emergency financial assistance to low-income countries hit by climate-induced disasters, among other things.[50] For example, in 2019 the RFI delivered $100 million to Mozambique a few weeks after Cyclone Idai killed 1300 people.[51] Second, re-establish a revenue base. Here consideration should be given to the changing nature of the global economy. Many countries embroiled in conflict tend to generate much of their revenue from the production and export of raw commodities, especially oil. However, major oil-consuming countries have announced policies to ban the sale of petrol and diesel cars from 2035, or earlier.[52] Therefore, allocating billions of dollars via multilateral funding mechanisms or direct donor support, to rebuild oil production and export facilities (ports and rail) may not yield a long-term sustainable revenue base, but rather result in stranded assets. A better long-term investment might be to develop a minerals extraction and export sector that is reoriented towards supporting global green growth aspirations. For example, cobalt and lithium-iron are essential for large-scale battery storage and are found in large quantities in conflict-affected and fragile states.[53]

The second, and much larger task, is for peacebuilders to begin to set the conditions for sustained economic growth.[54] This means developing and implementing a growth plan with local actors that embodies the following four features. First, directs investment and technical assistance towards low-emissions agriculture, energy, transport, and heavy industry.[55] Second, creates a taxation and regulatory system that prioritises low-emissions activities, for instance by establishing a fixed price on carbon emissions (which may become essential to access foreign markets such as the EU).[56] Third, develops a diplomatic service capable of promoting the country's green credentials abroad and works to attract foreign investment in low-emission activities.[57] Fourth, understands the risks posed by climate change to communities, businesses, and critical infrastructure, and invests in risk reduction and climate adaptation technologies, prioritising nature-based solutions.[58]

Social Programmes for Sustainable Peace

Providing support for social programmes can help to prevent the re-emergence of violence, resolve ongoing conflicts, and rebuild societies in the aftermath of an atrocity event.[59] Social programmes that deserve priority support include those that assist women become more central to major decision-making processes and those that facilitate the return of young people to school.[60] Climate change can feature in each of these traditional peacebuilding activities.

The role that women can and should play in climate-related peacebuilding activities has been in focus in recent years.[61] The most prominent example presented details the contribution that women and girls can make to 'water diplomacy' – defined as efforts to solve water-related conflicts and promote water cooperation for peace.[62] The UN estimates that at least one-quarter of the world's population will live in water-scarce regions by 2050.[63] Such projections have led to successive UN secretary-generals to link water scarcity to violent conflict.[64] Globally, women and girls are responsible for water collection in 80 per cent of households.[65] However, water diplomacy remains male-dominated.[66] Increased support for programmes that elevate the status of women in water negotiations, particularly those of a transboundary character, is considered to be of high importance by researchers and practitioners alike.[67] The first step is to support education and relevant employment opportunities, with clear pathways identified to leadership positions.[68]

Getting children back to school is the other critical social underpinning to achieve a sustainable peace because it signals return to normalcy and shows that the country is willing to invest in its future.[69] For these reasons, rebuilding schools, training teachers, and distributing materials must be given priority. Where possible, schools could provide information to students and the broader community about the projected impacts and risks associated with climate change at a local level, and how individuals and communities can prepare, react, and adapt.[70] Special attention should be paid to the mental health impacts of climate change on young people – commonly referred to as 'climate anxiety'. There are many international actors that can, and do, play a significant role in climate education and literacy in schools in fragile states. For example, UNESCO together with several UN agencies, especially UNICEF, have developed a range of online and freely available technical guidance material and teaching and learning resources for primary and secondary schools with titles such as 'Climate Change in the Classroom', and 'Disaster Risk Reduction in the School Curricula', among many others.[71] The UN also works directly with governments with specific needs. For example, in the aftermath of Cyclone Nargis, UNESCO worked with the Government of Myanmar to develop a variety of age-appropriate climate impact preparedness and response material for students with titles such as 'Let's be Prepared for Disasters: Activity Book' (for primary schools), and teaching aids such as 'Introduction to Disaster Risk Reduction: Training Module 1'.[72]

Notes

1 Gareth Evans, *The Responsibility to Protect: Ending Mass Atrocity Crimes Once and For All* (Washington, DC: Brookings Institution Press, 2008), p 148.
2 Ibid.
3 Ibid. Some legal measures are not included in the discussion.
4 Ibid., p 149.
5 Florian Krampe, Farah Hegazi, and Stacy D. Van Deveer, 'Sustaining Peace Through Better Resource Governance: Three Potential Mechanisms for Environmental Peacebuilding', *World Development*, Vol 144, 2021, p 2. See also Anaïs Dresse Itay Fischhendler Jonas Østergaard Nielsen Dimitrios Zikos, 'Environmental Peacebuilding: Towards a Theoretical Framework', *Cooperation and Conflict*, Vol 54, No 1, 2019, p 99–119; and Tobias Ide, 'The Impact of Environmental Cooperation on Peacemaking: Definitions, Mechanisms, and Empirical Evidence', *International Studies Review*, Vol 21, No 3, September 2019, p 327–346.
6 UNEP, 2009 in McKenzie F. Johnson, Luz A. Rodríguez, and Manuela Quijano Hoyos, 'Intrastate Environmental Peacebuilding: A Review of the Literature', *World Development*, Vol 137, 2021, p 2.

7 Ken Conca and Geoffrey D. Dabelk (Ed.), *Environmental Peacemaking* (Washington, DC: Johns Hopkins University Press and Woodrow Wilson Center Press, 2002).
8 'Afghanistan Post-Conflict Environmental Assessment', United Nations Environment Programme, 2003 in McKenzie F. Johnson, Luz A. Rodríguez, and Manuela Quijano Hoyos, 'Intrastate Environmental Peacebuilding: A Review of the Literature', *World Development*, Vol 137, 2021, p 2.
9 Tobias Ide, Carl Bruch, Alexander Carius, Ken Conca, Geoffrey D. Dabelko, Richard Matthew, and Erika Weinthal, 'The Past and Future(s) of Environmental Peacebuilding', *International Affairs*, Vol 97, No 1, 2021, p 15. See also, Florian Krampe, 'Toward Sustainable Peace: A New Research Agenda for Post-Conflict Natural Resource Management', *Global Environmental Politics*, Vol 17, No 4, 2017, p 1–8.
10 See for example, 'Six Grave Violations Against Children in Times of War How Children Have Become Frontline Targets in Armed Conflict', *UNICEF*, 26 August 2021.
11 Evans, *The Responsibility to Protect* (Brookings, 2008), p 151.
12 Former UN Under-Secretary of Peacekeeping Operations, Marrack Goulding, *Peacemonger* (Washington, DC: John Hopkins University Press, 2003), p 18.
13 Evans, *The Responsibility to Protect* (Brookings, 2008), p 120–121.
14 See for example, 'The Brahimi Report' which makes operational recommendations on these issues.
15 Goulding, *Peacemonger*, p 18.
16 Ibid., p 19.
17 Ibid., p 18.
18 Some of the examples to come have been informed by the integrated peacebuilding discussion found in Beatrice Mosell and Lukas Rüttinger, 'Linking Adaptation and Peacebuilding: Lessons Learned and the Way Forward', *Climate Diplomacy and Adelphi*, 2019.
19 See for example, Dennis Tänzler and Nikolas Scherer, 'Guidelines for Conflict-Sensitive Adaptation to Climate Change', *Adelphi* for the German Environment Agency, July 2018. Lukas Ruettinger, Antoine Morin, Annabelle Houdret, Dennis Taenzler, and Clementine Burnley, 'Water, Crisis and Climate Change Assessment Framework (WACCAF)', *Adelphi* for Initiative for Peacebuilding funded by the European Union, June 2011; Amanda Kron, Lukas Rüttinger, Benjamin Pohl, Susanne Wolfmaier, Albert Martinez Sequero, Molly Kellogg, Asif Zaidi, and Daniel Essay, 'Addressing Climate-Related Security Risks', *Adelphi* for the UN Environment Programme, November 2019. See also, Kirby Reiling and Cynthia Brady, 'Climate Change and Conflict: An Annex to The USAID Climate-Resilient Development Framework', *USAid*, February 2015; and 'From Conflict to Peacebuilding the Role of Natural Resources and the Environment', *UNEP*, 2009, p 19–28.
20 See for instance, 'Integrating Environment in Post-Conflict Needs Assessments', Guidance Note, *UN Environment Programme*, March 2009. Evans, *The Responsibility to Protect* (Brookings, 2008), p 152.
21 Evans, *The Responsibility to Protect* (Brookings, 2008), p 154.
22 Ibid., p 154–155.
23 Demobilisation, the second most relevant option, could involve reclassifying ex-combatants to form natural disaster preparedness and response teams with an overlap in skills already present.
24 Evans, *The Responsibility to Protect* (Brookings, 2008), p 155–156.

25 These reintegration opportunities are the outcome of blending information contained the following reports: 'Office of Transition Initiatives: Lessons Learned, Disarmament, Demobilization and Reintegration Programming', *USAid*, August 2010, p 1; 'Stabilisation Issues Note: Disarmament, Demobilisation and Reintegration', Stabilisation Unit – UK Government (with UKAid, the foreign office and defence ministry), p 3. And Mosello and Rüttinger, 'Linking Adaptation and Peacebuilding'; Janani Vivekananda, Martin Wall, Chitra Nagarajan, Florence Sylvestre, and Oli Brown, 'Shoring Up Stability: Addressing Climate and Fragility Risks in the Lake Chad Region', *Adelphi*, 2019.

26 Florian Krampe, 'Why United Nations Peace Operations Cannot Ignore Climate Change', *SIPRI*, 22 February 2021. See also, Agathe Sarfati, 'Toward an Environmental and Climate-Sensitive Approach to Protection in UN Peacekeeping Operations', *International Peace Institute*, October 2022. See also the countries where the UN supports DDR initiatives in practice: 'Disarmament, Demobilization and Reintegration', *UN Peacekeeping*, 2022.

27 Krampe, 'Why United Nations Peace Operations Cannot Ignore Climate Change'. See also, Sarfati, 'Toward an Environmental and Climate-Sensitive Approach to Protection in UN Peacekeeping Operations'.

28 'The Role of Natural Resources in Disarmament, Demobilization and Reintegration: Addressing Risks and Seizing Opportunities', *UNDP* and *UNEP*, 2013, p 36–47. See also, 'Integrated Disarmament Demobilisation and Reintegration Standards. Module 6.30: DDR and Natural Resources', UNDDR Resource Centre (accessed 2019).

29 Evans, *The Responsibility to Protect* (Brookings, 2008), p 156.

30 Ibid.

31 'SSR in a Nutshell: Manual for Introductory Training on Security Sector Reform', *DCAF* and *ISSAT*, 2012, p 24.

32 'Preparing for the Coming Storm: The Security Sector and the Environment', *DCAF*, 2021, p 1.

33 Ibid., p 3.

34 Evans, *The Responsibility to Protect* (Brookings, 2008), p 158.

35 Ibid., p 162.

36 Ibid.

37 Rim Turkmani, Marika Theros, and Sami Hadaya (Eds.), *Political Economy and Governance in Syria: Presentations from the Political Economy and Governance in Syria*, Conference Organised at London School of Economics, 3–4 December 2018.

38 Ashley South, '"Hybrid Governance" and the Politics of Legitimacy in the Myanmar Peace Process', *Journal of Contemporary Asia*, Vol 48, No 1, November 2017, p 50–66.

39 Richard B. Stewart, Benedict Kingsbury, and Bryce Rudyk, 'Climate Finance: Key Concepts and Ways Forward', Harvard project on International Climate Agreements, 2017. Harald Winkler and Navroz K. Dubash 'Who Determines Transformational Change in Development and Climate Finance?', *Climate Policy*, Vol 16, No 6, 2016, p 783–791.

40 Winkler and Dubash, 'Who Determines Transformational Change in Development and Climate Finance?', p 783–791.

41 Ibid., p 163–169.

42 Evans identifies three conditions (physical, economic, and social), which have been adapted and expanded to be climate-facing, see Evans, *The Responsibility to Protect* (Brookings, 2008), p 168–169.

43 'Global Report on Internal Displacement (GRID) 2021', Report, Internal Displacement Monitoring Centre, 2021, p 1 and 12. See also, 'Climate Disasters 'Caused More Internal Displacement Than War' in 2020', *The Guardian*, 20 May 2021.

44 'Climate Crisis Could Displace 1.2bn People by 2050, Report Warns', *The Guardian*, 9 September 2020. See also, 'Guidance on Protecting People from Disasters and Environmental Change through Planned Relocation', Brookings, Georgetown University, and UNHCR, 7 October 2015.

45 This may become an issue of legality, see for instance, 'Climate Refugees Can't Be Returned Home, Says Landmark UN Human Rights Ruling', *The Guardian*, 20 January 2020.

46 'Climate Change and the Future of Safe Returns', *UNHCR* and *PIK*, November 2020.

47 Alice Chang, 'Recommendations for the Biden Administration on Climate Migration', *New Security Beat*, 20 July 2021. 'Report on the Impact of Climate Change on Migration', *The White House*, October 2021.

48 Evans, *The Responsibility to Protect* (Brookings, 2008), p 169–170.

49 Ibid., p 170.

50 'Rapid Financing Instrument', *International Monetary Fund*, 4 October 2022.

51 Seán Nolan and Krishna Srinivasan, 'When Disaster Strikes: Preparing for Climate Change', *IMF Blog*, 26 June 2019.

52 'UK Expected to Ban Sale of New Petrol and Diesel Cars From 2030', *The Guardian* 2020; 'European Union Bans Sale of Petrol and Diesel Cars from 2035 to Boost Electric Vehicle Uptake', *ABC News*, October 2022; 'California to Ban the Sale of New Gasoline Cars', *The New York Times*, 24 August 2022; 'Canada to Ban Sale of New Fuel-Powered Cars and Light Trucks from 2035', *Reuters*, 30 June 2021.

53 'Putting Peace at the Centre of the "Green Recovery" Agenda', Briefing Note, *International Alert* (funded by UKAid), May 2021, p 1. See also, 'The Role of Critical Minerals in Clean Energy Transitions', *International Energy Agency*, 2021.

54 Ibid., p 170.

55 For an overview of mitigation solutions see *Project Drawdown* (accessed 2022); and 'Green MENA Network', *Planetary Security*, 27 May 2020. See also, 'Solar and Wind Cheapest Sources of Power in Most of the World', *Bloomberg*, 28 April 2020; 'What Is Green Steel and Why Does the World Need More of It?', *World Economic Forum,* 11 July 2022; Jörgen Sandström, Yvonne Leung, Joerg Hildebrandt, and Kate Martin, '6 Countries Taking Action to Solve Concrete's Emissions Problems', *World Economic Forum*, 4 August 2022.

56 'EU Agrees to The World's Largest Carbon Border Tax', *CNN*, 19 December 2022.

57 Ben L. Parr, *Australian Climate Policy and Diplomacy: The Transition Years – John Howard to Kevin Rudd*, PhD Thesis, The University of Melbourne, 2015; *Australian Climate Policy and Diplomacy: Government-Industry Discourses* (Julia Gillard to Tony Abbott) (Oxon: Routledge, 2020).

58 Wim Zwijnenburg, 'Nature-Based Solutions Vital to Mitigating Conflict-Linked Environmental Damage', *New Security Beat*, 18 February 2021.

59 Evans, *The Responsibility to Protect* (Brookings, 2008), p 171.

60 Ibid.

61 See for example, 'Integrating Gender and Security in Climate Adaptation: Principles for Success', *International Alert* (funded by UKAid), Boyer, Meijer, and Gilligan, 'Advancing Gender in the Environment: Exploring the Triple Nexus of Gender Inequality, State Fragility, and Climate Vulnerability', *IUCN* and *USAid*, August

2020. See also, Hannah Elisabeth Kurnoth, Beatrice Mosello, and Janna Greve, 'Stronger Together', *Climate Diplomacy*, 14 June 2021; Adriana Abdenur, Mayesha Alam, Adam Day, Cristal Downing, and Beatrice Mosello, 'How Can Climate Considerations Be Better Integrated into the Women, Peace, and Security Agenda?', *Climate Diplomacy*, 25 October 2021; and, 'Summary of the UN Secretary-General's Report on R2P, Prioritizing Prevention and Strengthening Response: Women and the Responsibility to Protect', *Global Centre for R2P*, 6 August 2020.

62 Benjamin Pohl, 'The Rise of Hydro-Diplomacy Strengthening Foreign Policy for Transboundary Waters', *Adelphi*, 2014; Adrien Detges and Tobias Ide, 'International Water Cooperation Opens the Door to Peacemaking', *New Security Beat*, 27 November 2018; Martina Klimes, 'Water Diplomacy: Facilitating Dialogues', *Stockholm International Water Institute*, July 2019; 'Building Stability: Iraq's Interprovincial Water Disputes', *Planetary Security Initiative*, 10 July 2020; Erika Weinthal and Neda Zawahri, 'Don't Politicize Water (Levant)', *Foreign Policy*, 17 September 2020; Benjamin Pohl, Sabine Blumstein, and Susanne Schmeier, 'Strengthening Water Diplomacy', *Climate Diplomacy*, 13 April 2021.

63 Marisa O. Ensor, 'No Peace without Water, No Water without Peace, and Neither without Women's Empowerment', *Climate Diplomacy*, 5 April 2021.

64 'Former National Leaders: Water a Global Security Issue', United Nations University, 20 March 2011.

65 Marisa O. Ensor, 'No Peace without Water, No Water without Peace, and Neither without Women's Empowerment', *Climate Diplomacy*, 5 April 2021.

66 Ibid.

67 Natasha Carmi, 'Untapped Potential: Woman, Leadership and Water Diplomacy in the Middle East', Presentation to the Geneva Water Hub (Wilson Centre Event), June 2021

68 Ibid.

69 Evans, *The Responsibility to Protect* (Brookings, 2008), p 171.

70 'Climate Change Education', *UNESCO* (accessed 2022).

71 Ibid. See also, 'Climate Change Starters Guidebook: As Issues Guide for Education Planners and Practitioners', *UNEP*, *UNESCO* and *WHO*, 2011; 'Climate Change Adaptation and Disaster Risk Reduction in the Education Sector: Resource Manual', *UNICEF*, November 2012.

72 'Let's Be Prepared for Disasters: Activity Book', *Government of Myanmar* and *UNESCO*; and 'Introduction to Disaster Risk Reduction: Training Module 1', *Government of Myanmar* and *UNESCO*, February 2010. See also, Lynne Bethke, 'Evaluation of the UNESCO's Myanmar Education Recovery Programme', *InterWorks LLC* (consultancy), July 2010.

Conclusion

This book argued that the climate action and Responsibility to Protect (R2P) agendas share a common cause (and mission): to protect vulnerable human populations from large-scale harm. It also suggested that this common cause should serve as a rallying cry to establish a commitment from both camps to share information and ideas and to work more closely in the challenging decades ahead. To substantiate the central argument, and to provoke a conversation between climate and R2P actors, the book examined and highlighted the historical, conceptual, and operational parallels between both agendas, and sought to identify linkages where they could be found.

Part I investigated the history of the climate action agenda and the R2P agenda.

Chapter 1 examined the history of human-made global warming and mass atrocities – the parallel problems that the book sought to address. It showed that since the 1850s, global temperatures and carbon dioxide concentrations in the atmosphere have risen in a somewhat linear fashion. It also showed that the frequency and scale of mass atrocity situations have oscillated over time, notably declining between 2000 and 2010; however, since the Arab Spring in 2011, many of the key atrocity indicators have move notably upwards. In short, both warming and atrocities are getting worse. The chapter also showed that climatic change, industrialisation, fossil fuels, economic power, war, and atrocities are deeply intertwined.

Chapter 2 traced the parallel histories of major international policy efforts to reduce emissions and prevent atrocities. It showed that since the end of the Cold War, there have been numerous solutions proposed and implemented to address the problem of emissions (and by implication, climate change) and conflict (and by implication, atrocities). When considering the findings of Chapter 1, little doubt remains that emission reduction policies and atrocity prevention policies have largely failed to achieve their core objective: to protect vulnerable human populations. The chapter also showed that over time the climate action and atrocity prevention agendas would converge and diverge at various times. That is, each agenda would include or exclude key issues, ideas,

DOI: 10.4324/9780367201364-10

and concerns, from the other at different historical points, and in different historical periods, between 1988 and 2022.

Chapter 3 began by detailing the West's recent efforts (2010–2022) to integrate the climate action agenda (e.g. policies and ideas) into the conflict prevention and response agenda. It provided evidence of an underlying closeness between the climate action and conflict prevention agendas based on a shared interest in human protection; but more explicitly showed that integration has evolved over time. Initially (pre-Paris), concerns about climate impacts were linked to defence agendas and to a lesser extent, foreign policy agendas; following this (post-Paris), the climate action agenda (encompassing climate impacts as well as mitigation and adaptation policies) was increasingly being 'integrated' into the broader conflict prevention and response agenda; and finally (Biden era), actionable policies were identified and cautious initial links made between climate impacts and the atrocity prevention and response agenda. Western governments also increasingly sought to integrate climate action issues into the UN Security Council between 2010 and 2022.

This analysis provided a springboard to begin to examine the relationship between the climate action agenda and the R2P agenda (the premier atrocity prevention and response framework globally). The first section examined some of the R2P's key conceptual debates and issues that hold relevance to climate change. This investigation was undertaken in the context of a discussion about the prospects and limitations of expanding the R2P's 'scope' and its 'watchlist factors'. It was found that the R2P scope should remain limited to the four-fold crimes: genocide, war crimes, crimes against humanity, and ethnic cleansing – and therefore, continue to exclude 'environmental emergencies'. However, the R2P watchlist factors should consider including 'climate vulnerability' alongside the existing fivefold factors: past atrocities, continuing tensions, coping mechanisms, external receptiveness, and leadership quality.

Part II continued to parallel the R2P and climate action agendas by laying out the R2P's operational framework (commonly known as the 'R2P toolbox') and situating a variety of discrete climate action policies and issues within it.

Chapter 4 examined the strategies and policy options available under the responsibility to prevent (R2Prevent), which encompass early warning systems as well as a multitude of structural and direct diplomatic, economic, legal, and security policy options. The aim of the R2Prevent, as expressed in Evans's 2008 publication, is to reduce the risk of an atrocity situation occurring in the future. The chapter showed that a multitude of climate action policies and issues similarly seek to provide a structural human protection role by reducing emissions, as well as a direct human protection role by adapting human and ecological systems to unavoidable warming.

Chapter 5 investigated the responsibility to react (R2React) phase. The aim here is to stop an imminent or occurring atrocity situation. To achieve this, the R2React details a range of policy options under the same four strategies as the R2Prevent, with one conspicuous extension: the use of coercive military force. Climate action policies and ideas, it was shown, can parallel the full range of

R2React policies, including its most extreme option. However, the notion of mounting an 'ecological intervention' in response to a situation where human populations do not face grave risk appears to have little practical utility at the present time. Nonetheless, it was shown that most climate action options situated under the R2React phase were, or could be, geared towards satisfying human protection needs.

Chapter 6 briefly examined the responsibility to rebuild (R2Rebuild). The R2Rebuild details a variety of policies situated under security, diplomatic, legal, and economic strategies and is considered the beginning of a new process of conflict prevention, but with greater emphasises placed on structural policy options compared to direct options. The aim is to protect populations from a relapse into violent conflict. Climate action policies can parallel many of the R2Rebuild options – apart from a small gap in 'achieving justice and reconciliation' (legal options). The climate options identified seek to protect populations from large-scale harm by way of developing economies and communities able to withstand the (re)occurrence of a sudden or slow onset climate impact (cyclones versus drought) and/or by contributing to the global emissions reduction agenda.

In March 2023, the Intergovernmental Panel on Climate Change (IPCC) released its Synthesis Report for its Sixth Assessment Cycle, providing a summary of the key findings from Working Group I, II, and III, as well as three Special Reports.[1] The Synthesis Report, therefore, presented a round-up of the scholarly research on global temperature rise, its principal causes, and its associated impacts and risks – as well as laying out possible future scenarios on warming, causes and impacts. It also identified a multitude of climate action policies and actors operating across different yet interlinked levels, such as the international and national level via the UNFCCC and nationally; the subnational level, encompassing cities, states/provinces, local government, and businesses; as well as those operating nationally and internationally in the key emitting sectors: agriculture, energy, transport, and heavy industry. This matrix of policy action was captured by the UN Secretary-General, António Guterres, who subsequently declared: 'This report is a clarion call to massively fast-track climate efforts by every country and every sector and on every timeframe. Our world needs climate action on all fronts: everything, everywhere, all at once.'[2] However, the Synthesis Report was silent on the role that conflict/atrocity prevention and response policies and actors could play. If the IPCC's key message really is 'act now, or it will be too late',[3] then building a genuine and meaningful relationship with these researchers and practitioners must surely be part of the mix.

Notes

1 Alix Kashdan and Kristyn Ostanek, 'The IPCC'S 6th Report Cycle Is Ending. Here's the Latest on The State of Global Climate Science', *UN Foundation*, 20 March 2023. 'Synthesis Report of The IPCC Sixth Assessment Report (AR6)', Longer Report,

IPCC, March 2023. Also see 'Synthesis Report of The IPCC Sixth Assessment Report (AR6)', Summary for Policymakers, *IPCC*, March 2023. See also, 'Scientists Deliver "Final Warning" on Climate Crisis: Act Now or It's Too Late', *The Guardian*, 21 March 2023.

2 'Secretary-General Calls on States to Tackle Climate Change "Time Bomb" through New Solidarity Pact, Acceleration Agenda, at Launch of Intergovernmental Panel Report', *Messages and Statements*, 20 March 2023.

3 'Scientists Deliver "Final Warning" on Climate Crisis: Act Now or It's Too Late', *The Guardian*, 21 March 2023.

Index

Note: Page numbers in **bold** refer to tables.

For Product Safety Concerns and Information please contact our EU representative GPSR@taylorandfrancis.com Taylor & Francis Verlag GmbH, Kaufingerstraße 24, 80331 München, Germany